FLORIDA'S
SEASHELLS
A Beachcomber's Guide

Blair and Dawn Witherington

Pineapple Press, Inc.
Sarasota, Florida

To our parents

Front Cover Photographs

Rose petal tellin *(Tellina lineata)*
Common jingle shell *(Anomia ephippium)*
Alphabet cone *(Conus spurius atlanticus)*
Turkey wing (zebra ark) *(Arca zebra)*
Banded tulip *(Fasciolaria lilium)*
Even pricklycockle *(Trachycardium isocardia)*
Zigzag scallop *(Euvola ziczac)*
Half-naked penshell *(Atrina seminuda)*
Globe purple sea snail *(Janthina prolongata)*

Front Flap

Frond Oyster *(Dendostrea frons)*

Acknowledgments

For their contributions, reviews, and advice we are greatly indebted to Dean Bagley, Michael Bresette, Bill Frank, and Inwater Research Group, Inc.

All photographs are © Blair Witherington and Dawn Witherington

Inquiries should be addressed to:

Pineapple Press, Inc.
P.O. Box 3889
Sarasota, Florida 34230

www.pineapplepress.com

Library of Congress Cataloging-in-Publication Data

Witherington Blair E., 1962-
  Florida's seashells: a beachcomber's guide / Blair and Dawn Witherington.
    p. cm.
  Includes index.
  ISBN 978-1-56164-387-5 (pbk. : alk. paper)
  1. Shells—Florida—Identification. I. Witherington, Dawn. II. Title.
  QL415.F6W56 2007
  594.147'709759—dc22

                    2007010729

First Edition
10 9 8 7 6 5 4 3 2

Design by Dawn Witherington
Printed in China

Contents

Introduction

Seashells on beaches satisfy the searcher in each of us. They have all it takes to trigger the collection compulsion—beauty, variety, mystery, intrigue, and a pocket-sized form. Seashells also happen to be scattered upon one of our favorite places, where a barefoot, sandy-seaside stroll can reveal a bounty of collectibles even to a casual visitor. And for a searcher with eyes sharpened by images of potential finds, there is a diverse world spiced with rare and provocative shells that most others pass by.

So what are these seashells? They are, or were, the protective and supportive parts of soft-bodied, marine mollusks—animals in the phylum Mollusca. For most shells found on beaches, only the animal's persistent parts remain after its softer bits have fed other elements of the food chain. Shells persist on beaches due to their mineral makeup—calcium carbonate crystals laid down in opposing directional layers held together by small amounts of protein glue. Of course, this chemical description does nothing to explain the persistence of seashells within the long history of the human experience. Only the natural poetry of a shell's exquisite form can do that.

Four classes of mollusk shells are represented in this book:

 Gastropods (class Gastropoda, meaning "stomach-footed") are symmetrical animals twisted into an asymmetrical shape—the familiar spirally coiled snails.

 Cephalopods (class Cephalopoda, meaning "head-footed") have distinct heads, complex eyes, and a set of armlike tentacles. For most, including octopi, cuttlefish, and squid, the shell is internal.

 Scaphopods (class Scaphopoda, meaning "boat-footed") are eyeless animals that live in tusklike shells that are open at each end.

 Bivalves (class Bivalvia, meaning "two-shelled") have valves (shells) connected by a hinge and include the familiar oysters, scallops, and clams.

Finding Beach Shells

The most ardent beachcombers know when to look for shells—every chance they get. But folks with busy lives may want some temporal guidance. Hot times for seashell finds include late winter and spring, especially in southwest Florida. These are times when deep swells move seashells onto beaches. Late spring storms bring great shelling on the Atlantic coast, and hurricanes (summer and fall) can scatter shells onto any of

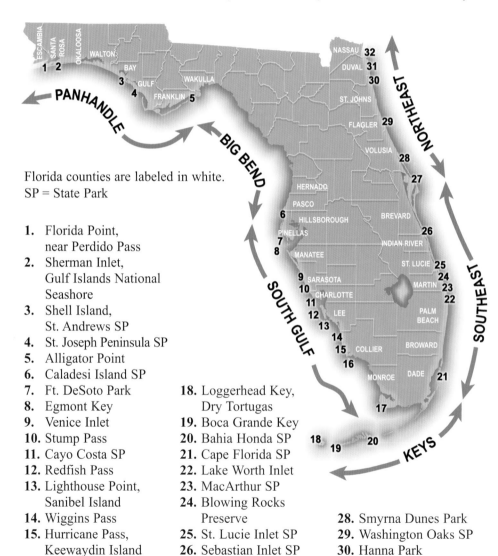

Florida counties are labeled in white.
SP = State Park

1. Florida Point, near Perdido Pass
2. Sherman Inlet, Gulf Islands National Seashore
3. Shell Island, St. Andrews SP
4. St. Joseph Peninsula SP
5. Alligator Point
6. Caladesi Island SP
7. Ft. DeSoto Park
8. Egmont Key
9. Venice Inlet
10. Stump Pass
11. Cayo Costa SP
12. Redfish Pass
13. Lighthouse Point, Sanibel Island
14. Wiggins Pass
15. Hurricane Pass, Keewaydin Island
16. Cape Romano
17. Cape Sabal

18. Loggerhead Key, Dry Tortugas
19. Boca Grande Key
20. Bahia Honda SP
21. Cape Florida SP
22. Lake Worth Inlet
23. MacArthur SP
24. Blowing Rocks Preserve
25. St. Lucie Inlet SP
26. Sebastian Inlet SP
27. Canaveral National Seashore

28. Smyrna Dunes Park
29. Washington Oaks SP
30. Hanna Park
31. Little Talbot Island SP
32. Fort Clinch SP

the state's beaches. Some of the best seashell abundance and diversity occurs on beaches during the calm days following rough weather. On any day of the year, the best shelling is at low tide. Shelling is especially good near the full- and new-moon periods of the month when tides are lowest.

Where on the beach to look for seashells depends upon the kind of shells one seeks and how the ocean has recently treated the beach. Combing the wet sand at low tide is the best way to find small and delicate seashells, especially when the surf is calm. Calm seas may allow shellers to search the surf itself, including the highly productive drop-off (step) just seaward from the low-tide line. The recent high-tide line at mid-beach is often a good place to find large or fluttery shells. Keep in mind that the high-tide mark (the wrack line) from previous days may have been higher up the beach, where shells can be found if they have not been covered with wind-blown sand. The largest waves during the highest tides sweep up the beach to the storm wrack, which is often at the base of the dune. This area can hold large shells, and on infrequently combed beaches this upper beach is filled with rare finds.

No single beach fully represents all the seashells that Florida has to offer. Shells that are common on some beaches are rare on others. Beachcombers who seek a diverse collection, or merely a diverse experience, will want to visit a wide array of beaches. The list on the facing page is a selection of some of the top seashell spots that Florida has to offer. These top spots include beaches near inlets, at the tips of islands, fronting offshore reefs, and within the landward sweep of coastal currents.

Guide Organization

Each seashell in this book has a **range map** showing where in Florida one might find it. These ranges pertain specifically to an item's beach distribution, which may be different from the places occupied by the living mollusks. Coastal lines on the maps are solid where an item is relatively common, and dotted where relatively uncommon. Because the range maps are not

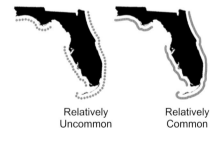

Relatively Uncommon   Relatively Common

absolute, a gap may indicate either rarity or uncertainty. All maps show a gap along Florida's Big Bend, where the marshy, submerged coastline has little or no sandy beach for beachcombing. The shell **sizes** given refer to maximum length unless otherwise indicated.

Because this is a guide to beach-found seashells, the depictions that follow are of beached shells. That is, many are likely to show the characteristic wear resulting from a little surf-tumbling. This beach-wear often creates forms that look different from the museum specimens portrayed in most shell books. Although most shells on beaches are dead, many mollusks live their lives in the surf zone. **Numerous Florida beaches prohibit the taking of shells still in use by their inhabitants.**

Shelled Mollusk Anatomy

Seashells are the protective or supportive skeletons of mollusks (phylum Mollusca). The most common shells are from snails (gastropods, with one coiled shell) and bivalves (with two hinged shells). Other shelled mollusks include tusk shells and some squids.

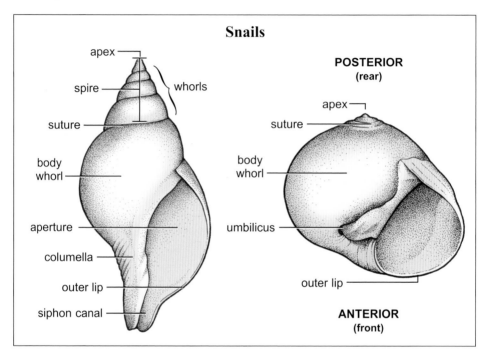

Snails

apex

spire

whorls

suture

body whorl

aperture

columella

outer lip

siphon canal

POSTERIOR (rear)

apex

suture

body whorl

umbilicus

outer lip

ANTERIOR (front)

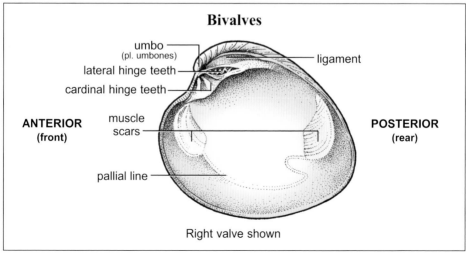

Bivalves

umbo (pl. umbones)

ligament

lateral hinge teeth

cardinal hinge teeth

ANTERIOR (front)

muscle scars

POSTERIOR (rear)

pallial line

Right valve shown

Limpets and Falselimpet

Cayenne
Keyhole Limpet

Cancellate
Fleshy
Limpet

Barbados and Rosy
Keyhole Limpets and
Striped Falselimpet

Cayenne keyhole limpet, max 2 in (5 cm)

RELATIVES: Keyhole limpets (family Fissurellidae) are only distantly related to falselimpets (family Siphonariidae).

IDENTIFYING FEATURES: All are shaped like oval-based volcanoes.

Cayenne keyhole limpets *(Diodora cayenensis)* have a keyhole opening offset from center.

Cancellate fleshy limpet, max 1.5 in (3.8 cm)

Cancellate fleshy limpets *(Lucapina suffusa)* are flattened with an oval top hole.

Barbados keyhole limpets *(Fissurella barbadensis)* have a nearly central, oval hole, and alternating small and large radiating ribs.

Barbados keyhole limpet, max 1 in (2.5 cm)

Rosy keyhole limpets *(Fissurella rosea)* have a broad oval top hole and pink or purple rays.

Striped falselimpets *(Siphonaria pectinata)* have a peak with no top hole.

HABITAT: On and under rocks. Falselimpets are frequently above water.

Rosy keyhole limpet, max 1.5 in (3.8 cm)

DID YOU KNOW? A limpet moves about by rippling its muscular foot and feeds on algae using a rasping, tongue-like radula. They anticipate low tide and return to resting spots before the water leaves their rock. Their top hole passes exhaust from their gills. Holeless falselimpets breathe air and can be found above the tide on jetty rocks.

Striped falselimpet, max 1 in (2.5 cm)

3

Scuptured topsnail, max 1 in (2.5 cm)

Jujube topsnail, max 1 in (2.5 cm)

Smooth Atlantic tegula, max 3/4 in (19 mm)

Topsnails and Tegula

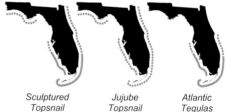

| Sculptured Topsnail | Jujube Topsnail | Atlantic Tegulas |

RELATIVES: Topsnails and tegulas are in the family Trochidae.

IDENTIFYING FEATURES:

Sculptured topsnails *(Calliostoma euglyptum)* have a dark-tipped apex and rounded whorls.

Jujube topsnails *(Calliostoma jujubinum)* are mottled red-brown or tan with a straight-sided, conelike shape.

Smooth (silky) Atlantic tegulas *(Tegula fasciata)* have a round-top turban shape.

HABITAT: Topsnails live on reefs and hardbottom. Smooth Atlantic tegulas prefer shallow seagrass beds.

DID YOU KNOW? Jujube topsnails owe their name to the jujube fruit (Chinese date, *Zizyphus jujuba*), which only remotely approximates the shape and color of a topsnail. Tegula is Latin for "roof tile" and refers to how the snail's spire resembles the shape and surface of a tiled roof. Each of these snails feeds on algae and detritus.

Star Shells and Chestnut Turban

Star Shell   Chestnut Turban

RELATIVES: Star and turban shells are in the family Turbinidae.

IDENTIFYING FEATURES: All are coarsely sculptured. Their colors are typically yellow to brown, but many beached shells are bleached white.

Long-spined star shells *(Astralium phoebium)* have saw-toothed spines at the whorl sutures.

American star shells *(Lithopoma americanum)* have a conelike spire with lumps instead of spines.

Chestnut turbans *(Turbo castanea)* have rounded whorls like a beaded turban and have a circular aperture.

HABITAT: On and under rocks in shallow hardbottom areas.

DID YOU KNOW? Living American star shells are covered with a fluffy coat of algae that serves to camouflage them. Chestnut turbans are one of the most important grazers on algae in seagrass beds. The door to their aperture, their operculum, has a pearly sheen.

Long-spined star shell, max 1.75 in (4.5 cm)

American star shell, max 1 in (2.5 cm)

Chestnut turban, max 1.5 in (4 cm)

Four-tooth nerite, max 1 in (2.5 cm)

Antillean nerite, max 1 in (2.5 cm)

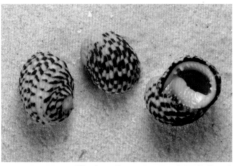

Checkered nerite, max 1 in (2.5 cm)

Virgin nerite, max 1/2 in (12.7 mm)

Nerites

Four-tooth, Antillean, and
Checkered Nerites

Virgin Nerite

RELATIVES: These "baby's tooth shells" are in the family Neritidae.

IDENTIFYING FEATURES: All have rounded spires and apertures that look like a toddler's smile.

Four-tooth nerites *(Nerita versicolor)* have thick spiral ridges with blurred markings of black, greenish-white, and maroon. They have four teeth on the inside lip.

Antillean nerites *(Nerita fulgurans)* are streaked and have spiral ridges separated by light sutures. They have two small central teeth on the inside lip.

Checkered nerites *(Nerita tessellata)* have alternating black and white markings and two tiny teeth on the inside lip.

Virgin nerites *(Vitta virginea)* are glossy and variably patterned with waves and swooshes.

HABITAT: Four-tooth, Antillean, and checkered nerites live on and under rocks in intertidal areas. Virgin nerites live in shallow waters of bay margins and seagrass beds.

DID YOU KNOW? Nerites have a thick, calcified operculum (aperture door) that keeps them locked up tight and protected from desiccating heat when exposed at low tide. They cluster in groups to graze at night on algae covering intertidal rocks.

Common Sundial and Wentletraps

Common Sundial   Angulate, Humphrey's, and
Brown-banded Wentletraps

Common sundial, max 2.5 in (6.4 cm)

RELATIVES: Sundials are in the family Architectonicidae. Wentletraps are in the family Epitoniidae.

IDENTIFYING FEATURES:

Common sundials *(Architectonica nobilis)* have a deep umbilicus and a spire like a flattened cone.

Angulate wentletraps *(Epitonium angulatum),* like most wentletraps, have a rounded, thick-lipped aperture and distinct, widely spaced ribs. The body whorl has 9–10 thin ribs that are angled at the whorl shoulders.

Angulate wentletrap, max 1 in (2.5 cm)

Humphrey's wentletraps *(Epitonium humphreysii)* have 8–9 thick, rounded ribs on the body whorl.

Brown-banded wentletraps *(Epitonium rupicola)* have rounded ribs of varying strengths and spiral bands of white, tan, and brown.

HABITAT: Sundials and wentletraps live in sandy areas to moderate depths.

Humphrey's wentletrap, max 1 in (2.5 cm)

DID YOU KNOW? Sundials spend their days buried spire-down in the sand and emerge at night to feed on sea pansies. Angulate wentletraps get away with chewing chunks off living anemones by soothing them with a purple anesthetic. More than 20 wentletrap species are known in Florida. "Wentletrap" comes from *wendeltreppe,* German for a winding staircase.

Brown-banded wentletrap, max 3/4 in (20 mm)

7

Interrupted periwinkle, max 1/2 in (13 mm)

Marsh periwinkle, max 1 in (2.5 cm)

Mangrove periwinkle, max 1 in (2.5 cm)

Beaded periwinkle, max 3/4 in (20 mm)

Periwinkles

| Interrupted Periwinkle | Mangrove and Beaded Periwinkles | Marsh Periwinkle |

RELATIVES: Periwinkles are in the family Littorinidae.

IDENTIFYING FEATURES: Periwinkles have rounded apertures and conical spires.

Interrupted periwinkles *(Nodilittorina interrupta)* have white and purple-brown wavy lines.

Marsh periwinkles *(Littoraria irrorata)* have thick aperture lips and are patterned with dashed streaks on their spiral ridges.

Mangrove periwinkles *(Littoraria angulifera)* have thin shells, sharp aperture lips, and a groove in their lower columella/inner aperture.

Beaded periwinkles *(Cenchritis muricatus)* are blue-gray or pink-gray and covered with white, beadlike knobs.

HABITAT: Periwinkles live just above the tide. Marsh periwinkles live on marsh reeds, mangrove periwinkles live on mangrove shorelines, and both interrupted and beaded periwinkles live on rocks near wave splash.

DID YOU KNOW? Periwinkles feed out of the water on algae that grows on plants and rocks, and their beached shells are most common near inlets. These snails are an important link in the food chain between estuarine plants and dozens of crab, fish, and bird species.

Worm Shells

Variable Worm Shell

Black Worm Shell and Corroding Worm Shell

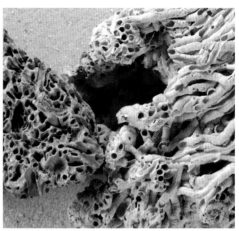

Variable worm shell

RELATIVES: Worm shells are in the family Vermetidae and are distantly related to ceriths and worm snails.

IDENTIFYING FEATURES: Worm shells are snails that grow in irregular patterns.

Variable worm shells *(Petaloconchus varians)* have an aperture to 1/16 in (1.6 mm) and grow in compact, rocklike colonies of tangled tubes. Chunks of colonies are orange, brownish-purple, or whitish.

Black worm shells *(Petaloconchus nigricans)* have apertures to 3/16 in (5 mm) with horizontally wrinkled, purple-black shells. They grow in masses on hard surfaces between the tide lines.

Corroding worm shells *(Dendropoma corrodens)* have apertures to 1/4 in (6 mm) and are lumpy and white outside, shiny brown inside. They attach as individuals or groups to rocks and shells.

HABITAT: Shallow reefs, hardbottom, and exposed rock.

DID YOU KNOW? These snails let their surroundings dictate their worm-like form. Some anchor to the bottom and form reefs. Worm-shell gastropods feed in place by gill-filtering plankton. Some south Florida mangrove islands owe their origins to reef-building variable worm shells.

Black worm shell

Corroding worm shell

9

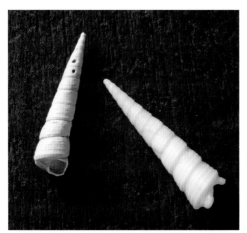

Boring turretsnail, max 1.5 in (3.8 cm)

West Indian wormsnail

Florida wormsnail

Turretsnail and Wormsnails

Boring Turretsnail and
Florida Wormsnail

West Indian Wormsnail

RELATIVES: Turrets and wormsnails are in the family Turritellidae.

IDENTIFYING FEATURES:

Boring turretsnails *(Turritella acropora)* have pale, sharp spirals with rounder whorls than the similar-shaped auger shells. Turret apertures have no siphon canal.

West Indian wormsnails *(Vermicularia fargoi)* are more loosely coiled than turrets and have three spiral cords leading to a squarish aperture. Older West Indian wormsnails abandon the coil theme after reaching 1 in (2.5 cm) long and begin growing freestyle.

Florida wormsnails *(V. knorrii)* grow wormlike after reaching about 1/2 in (1.25 cm). Both wormsnails are brownish, but the Florida wormsnail often has a white spiral tip.

HABITAT: Boring turrets live in shallow sandy bottom. Wormsnails grow with sponges on reefs and hardbottom.

DID YOU KNOW? Wormsnails live life attached to the bottom or to other wormsnails and feed on suspended plankton and detritus. Their immediate environment dictates their uncoiled growth. Some wormsnails literally tie themselves together in knots.

Ceriths

Florida Cerith   Stocky, Fly-specked,
                        and Dwarf Ceriths

RELATIVES: Ceriths are in the family Cerithiidae and are distantly related to worm shells and turretsnails.

IDENTIFYING FEATURES: Ceriths *(Cerithium* spp.) are bead-sculptured, slender snails with many whorls and distinct siphon canals opposite their pointed spires.

Florida ceriths *(C. atratum)*, 1.5 in (3.5 cm), have 18–20 beaded ridges per whorl and occasional larger lumps. Their beached shells are light to dark, often spiraled with brown and white.

Stocky ceriths *(C. litteratum)* are compact and, compared to Florida ceriths, have fewer but larger beads in their sculpture.

Fly-specked ceriths *(C. muscarum)* have 9–11 ridges per whorl that are crossed by spiral lines. New shells are "fly-specked" with spiral rows of brown dots.

Dwarf Atlantic ceriths *(C. lutosum)* are brown with a light apex and a thick aperture lip.

HABITAT: All live on subtidal sandy bottom or seagrass.

DID YOU KNOW? Ceriths feed on algae and detritus. The Florida cerith is the most common beached species, but all can be abundant near inlets.

Florida cerith, max 1.5 in (3.5 cm)

Stocky cerith, max 1 in (2.5 cm)

Fly-specked cerith, max 1 in (2.5 cm)

Dwarf cerith, max 3/4 in (20 mm)

West Indian false cerith, max 3/4 in (20 mm)

Ladder horn snail, max 1.2 in (3 cm)

Button snail, max 1/2 in (12.7 mm)

False Cerith, Ladder Horn Snail, and **Button Snail**

West Indian False Cerith and Ladder Horn Snail   *Button Snail*

RELATIVES: False ceriths (family Batillariidae), hornsnails (Potamididae), and button snails (family Modulidae) are all distantly related to ceriths.

IDENTIFYING FEATURES:

West Indian false ceriths *(Batillaria minima)* are gray to black with light spiral bands. Their aperture is brown with a short siphon canal twisting to the left.

Ladder hornsnails *(Cerithidea scalariformis)* have regularly spaced ribs, light spiral bands, and a flared aperture lip.

Button snails *(Modulus modulus)* are buttonlike with a gray- or brown-streaked, ridge-sculptured body whorl and a low spire.

HABITAT: West Indian false ceriths are locally abundant in shallow lagoons with coral rubble. Ladder hornsnails prefer mangrove mudflats and button snails like shallow seagrass.

DID YOU KNOW? West Indian false ceriths are a major food item for Caribbean flamingos. All three species feed on algae and detritus and deposit their eggs in gelatinous strings. The young hatch as free-swimming larvae, except for the button snail, which emerges as a miniature version of the adult.

Purple Sea Snails

Common purple sea snail, max 1.5 in (3.5 cm)

RELATIVES: Purple sea snails *(Janthina* spp.) are in the family Janthinidae, distantly related to wentletraps.

IDENTIFYING FEATURES: These gastropods have fragile violet shells. Live snails have a translucent, bubble-raft arcing from their aperture.

Common purple sea snails *(J. janthina)* have a low spire and D-shaped aperture. Their top whorls are pale and their base is violet.

Globe purple sea snails *(J. prolongata)* are all violet and have rounded whorls and a pointed spire.

HABITAT: Purple sea snails live adrift on the open ocean. Unbroken snails are found in freshly beached wrack.

DID YOU KNOW? Two other species are known from Florida beaches. Dwarf purple sea snails *(J. globosa)* have a deeply indented outer aperture, and pale purple sea snails *(J. pallida)* have large, round apertures. Purple sea snails sail along with, and prey upon, Portuguese man-o-war and by-the-wind sailors. When they are not attached to these floating hydrozoans, they construct a mucous-bubble raft for buoyancy. The snails' violet shell-tinting blends in with the color of deep ocean waters and pre-sumably hides them from the birds and young sea turtles that eat them. Ancient Greeks used a violet fluid produced by *Janthina* snails as a dye for clothing.

Common purple sea snail with bubble raft

Globe purple sea snail, max 3/4 in (20 mm)

13

Florida fighting conch, max 4 in (10 cm)

Hawk-wing conch, max 4 in (10 cm)

Juveniles: fighting conch (A), hawk-wing conch (B)

Conchs
(Florida Fighting and Hawk-wing)

Florida Fighting Conch   Hawk-Wing Conch

RELATIVES: These are true conchs in the family Strombidae.

IDENTIFYING FEATURES:

Florida fighting conchs *(Strombus alatus)* are thick-shelled with blunt-knobbed whorls. Colors vary from pale yellow to chestnut-brown with occasional light spots and zigzags. The body whorl has fine spiral cords (ridges).

Hawk-wing conchs *(Strombus rainus)* have coarse spiral cords on the body whorl and a rear projection to their widely flared aperture lip. The rough upper part of the "hawk wing" is mottled brown and white.

HABITAT: Florida fighting conchs live in sandy shallows, including the swash zone of low-energy beaches. Hawk-wing conchs prefer seagrass pastures.

DID YOU KNOW? Fighting conchs (pronounced "konks") get their name from occasional bouts between rival males. They are spry for snails and can quickly flip themselves and walk using their pointed operculum. Both species feed on algae and detritus. A similar southeast Florida and Caribbean species, the West Indian fighting conch *(S. pugilis),* has longer spines on the spire and is more orange. Fighting conchs are being farmed experimentally as an edible alternative to the rarer and slower-maturing queen conch.

Queen Conch

Queen conch, max 12 in (30 cm)

RELATIVES: This is a true conch in the family Strombidae.

IDENTIFYING FEATURES:

Queen conchs *(Strombus gigas)* are also known as pink conchs for their interior color, a hint of which can be seen on the exterior of fresh shells. Adults have a wide, thick, flaring aperture lip. Knobs on their whorl shoulders are especially pointed on the last three whorls. Juveniles have a more strongly pointed spire, a narrow, pointed base, and a sharp, unflared lip.

HABITAT: Queen conchs live in seagrass and coral rubble to moderate depths.

DID YOU KNOW? Florida's queen conchs have been nearly lost. Over-harvesting for meat and souvenirs has made this conch a rare beach find except for bleached old shells. Conch meat from the Bahamas continues to be sold as Keys cuisine and Florida shell shops still stock Bahamian conch shells. Live queen conchs in Florida have been protected from harvest since 1985, but recovery has been limited by poor water quality surrounding the Keys. Adult conchs are active at night when they may make journeys of more than a hundred yards (about 100 m) before sunrise. Queen conchs reach adulthood in about 4 years. Those in deep water where harvest pressure is low can reach 26 years of age.

Juvenile queen conch (size 6 in, 15 cm)

Beach-worn adult

15

Atlantic slippersnail, max 2.5 in (6.5 cm)

Spotted slippersnail, max 1 in (2.5 cm)

Spiny slippersnail, max 1 in (2.5 cm)

Eastern white slippersnail, max 1 in (2.5 cm)

White hoofsnail, max 3/4 in (2 cm)

Slippersnails and Hoofsnail

White Hoofsnail   *Slippersnails*

RELATIVES: Hoofsnails (family Hipponicidae) are only distantly related to slippersnails (family Calyptraeidae).

IDENTIFYING FEATURES: Slippersnails *(Crepidula* spp.) are shoe-shaped, with a conspicuous ventral shelf.

Atlantic slippersnails *(C. fornicata)* have a coiled apex bent to one side, a smooth exterior, and a shelf with an indented edge.

Spotted slippersnails *(C. maculosa)* have a shelf with a straight edge angling away from the apex. Most spotted slippersnails have brown spots on white.

Spiny slippersnails *(C. aculeata)* differ in having roughened, sometimes spiny, radiating ridges.

Eastern white slippersnails *(C. atrasolea)* are white, thin, and flattened with a small pointy apex.

White hoofsnails *(Hipponix antiquatus)* are white, roughened, thick-shelled, and shaped like a floppy Santa hat.

HABITAT: Hoofsnails attach themselves to rocks in areas of moving water. Slippershells live in shallow waters on rocks and on other shells. White slippersnails prefer to be inside other shells.

DID YOU KNOW? Slippersnails begin life as males that grow into being female. Environmental conditions determine when they strategically switch sex.

Cowries and Trivias

Cowries and
Fourspot Trivia

Coffeebean Trivia

Atlantic deer cowrie, max 5 in (13 cm)

Measled cowrie, max 4.5 in (11 cm)

RELATIVES: Cowries (family Cypraeidae) are distantly related to trivias (family Triviidae).

IDENTIFYING FEATURES: Cowries have glossy, egg-shaped shells with a body-length, grinning aperture. Florida's trivias are cowrie-shaped with riblets wrapped between a back groove and the aperture.

Atlantic deer cowries *(Macrocypraea cervus)* are chocolate-brown with solid white spots or hazy brown with light bands.

Measled cowries *(Macrocypraea zebra)* are similar to deer cowries, but are more elongate (less domed) and have dark centers to their side spots.

Atlantic yellow cowrie, max 1.2 in (3 cm)

Atlantic yellow cowries *(Erosaria acicularis)* have a granular yellow pattern with marginal brown spots.

Coffeebean trivias *(Niveria pediculus)* are pale purple with 3 pairs of dark spots.

Fourspot trivias *(Niveria quadripunctata)* are pinkish-white with 2–4 dark spots straddling the back groove.

Coffeebean trivia, max 3/4 in (20 mm)

HABITAT: Cowries and trivias live on shallow reefs.

DID YOU KNOW? Trivias feed on tunicates and soft corals. Cowries feed on algae and colonial invertebrates. The Atlantic deer cowrie is the largest of the world's 190 cowrie species.

Fourspot trivia, max 3/8 in (10 mm)

17

Colorful moonsnail, max 2 in (5 cm)

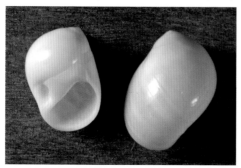

Milk moonsnail, max 1.5 in (4 cm)

White baby's ear, max 2 in (5 cm)

Maculated baby's ear, max 1.2 in (3 cm)

Moonsnails and Baby's Ears

| Colorful and Milk Moonsnails | White Baby's Ear | Maculated Baby's Ear |

RELATIVES: Moonsnails and baby's ears share the family Naticidae with shark's eyes and naticas.

IDENTIFYING FEATURES: All have a large body whorl, gaping aperture, and low, smooth spire.

Colorful moonsnails *(Naticarius canrena)* have a deep umbilicus half-filled with a traguslike pad (callus). They are creamy white with brown zigzags that are faded in old beach shells.

Milk moonsnails *(Polinices lacteus)* have a deep umbilicus half-filled with callus. Shells are glossy white.

White baby's ears *(Sinum perspectivum)* are like a flattened moonsnail with an expansive aperture. The body whorl is sculptured with broad spiral grooves. Shells are dull white or stained.

Maculated baby's ears *(Sinum maculatum)* are like the white baby's ear but are smudged with brown and have a higher, slightly pointed spire.

HABITAT: Sandy shallows.

DID YOU KNOW? Colorful moonsnails fade after being beached, but in life they have a lovely patterned shell and an enormous brown-streaked foot spreading ten times their shell size. Baby's ear snails have an equally big foot that cannot be withdrawn. These species prey on buried bivalves.

Shark's Eye

Shark's eye, max 3 in (7.5 cm)

RELATIVES: Shark's eyes (family Naticidae) are related to naticas, moonsnails, and baby's ears.

IDENTIFYING FEATURES:

Shark's eyes *(Neverita duplicata)* have a gaping aperture and a large body whorl that forms a smooth dome with their low spire. The umbilicus is nearly covered by a brown, traguslike pad (callus). Shells are brown-gray, blue-gray, or faded. Unfaded shells have a blue "eye" in the early whorls. Shark's eyes from the southern Gulf coast are browner with conelike spires and may be a distinct species.

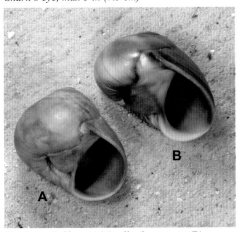

Southern Gulf coast (A), all other coasts (B)

HABITAT: Sandy shallows, including the swash zone.

DID YOU KNOW? Shark's eyes plow through surf-zone sands in search of clams. Unlucky clams are enveloped by the snail's foot while an acidic secretion softens the clam shell and a tooth-studded tongue (radula) rasps a beveled hole. The hole allows a visit from a proboscis that injects digesting enzymes and later removes liquid clam. Shark's eyes breed in the surf zone by cementing their eggs with sand into a gelatinous sheet that cures into a rubbery **sand collar**, which is hydrodynamically engineered to remain upright on shifting surf sands. Collars disintegrate when eggs hatch, so whole collars found in the swash contain developing little snails.

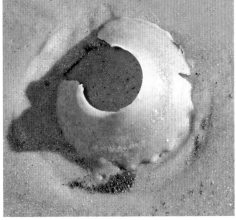

A sand collar in the swash zone

Scotch bonnet, max 4 in (10 cm)

Queen helmet, max 12 in (30 cm)

Reticulate cowrie-helmet, max 3 in (7.6 cm)

Scotch Bonnet and Helmet Shells

Scotch Bonnet   *Queen Helmet and Reticulated Cowrie-helmet*

RELATIVES: These gastropods are in the family Cassidae, distantly related to tun and fig shells.

IDENTIFYING FEATURES: All have a large body whorl with wide, toothy, grinning (or smirking) apertures.

Scotch bonnets *(Phalium granulatum)* are light shells with spiral grooves and a pointed spire. Colors range from white to cream with dark squares.

Queen (cameo) helmets *(Cassis madagascariensis)* are heavy shells with a low spire and a glossy, triangular aperture shield. Their whorl shoulders are studded with blunt knobs. Aperture teeth are cream on chestnut. The remaining shield is light or salmon with no central dark blotch like the king helmet *(Cassis tuberosa,* Keys only).

Reticulate cowrie-helmets *(Cypraecassis testiculus)* are dense, egg-shaped shells with smooth spiral grooves and growth lines, and a rounded spire. They are chestnut to salmon with darker, blurry squares.

HABITAT: Scotch bonnets occur in sandy shallows. Queen helmets and cowrie-helmets live on rocky reefs.

DID YOU KNOW? These gastropods feed on sand dollars and sea urchins. Queen helmet populations are low and declining, perhaps due to trawling and habitat loss.

Tun Shells and Figsnail

Giant Tun Shell and Atlantic Figsnail   *Atlantic Partridge Tun*

Giant tun shell, max 10 in (25 cm)

RELATIVES: Tun shells (family Tonnidae) are distantly related to figsnails (family Ficadae).

IDENTIFYING FEATURES: All are fragile, spiral-ridged shells with a large body whorl and a low spire.

Giant tun shells *(Tonna galea)* are almost spherical in shape with a wide aperture, prominent spiral ridges, and a plain cream or brown color. Most beach finds are in pieces.

Atlantic partridge tuns *(Tonna maculosa)* are more elongate than the giant tun, with a blurry pattern of alternating cream and brown rectangles.

Atlantic figsnails *(Ficus papyratia)* are delicately tapered at the front, have a low spire, and are sculptured with fine spiral ridges. Their colors range from cream to tan, sometimes with faint brown dots.

Atlantic partridge tun, max 5 in (13 cm)

HABITAT: Giant tuns are most common offshore. Atlantic partridge tuns and figsnails live in sandy shallows.

DID YOU KNOW? Tun shells feed on other mollusks, sea cucumbers, and fishes by engulfing their prey within a large expandable proboscis. Figsnails feed on sea urchins, and in life their shells are covered by their large, soft mantle (the part of their body that produces the shell).

Atlantic figsnail, max 5 in (13 cm)

21

Atlantic trumpet triton, max 12 in (30 cm)

Atlantic hairy triton, max 5 in (13 cm)

Giant hairy triton, max 3.5 in (9 cm)

Triton Shells

Atlantic Hairy and Trumpet Tritons

Giant Hairy Triton

RELATIVES: Tritons are in the family Ranellidae.

IDENTIFYING FEATURES:

Atlantic trumpet tritons (*Charonia variegata*) have a sharply spired, cone-shaped shell sculpted with wide spiral bands. The aperture has white teeth on chestnut-colored lips. Colors include rows of brown smears on a tan background.

Atlantic hairy tritons (*Cymatium aquatile*) have an elongate oval shell sculpted with spiral bands and strong axial ridges, 2/3 of a whorl apart, which represent former outer aperture lips. The current outer lip is also thick and bears paired whitish teeth. The inner lip has evenly spaced teeth.

Giant hairy tritons (*Cymatium parthenopeum*) have a thick, wavy outer lip.

HABITAT: Shallow to deep reefs.

DID YOU KNOW? Rome's Trevi Fountain features the Greek merman-god Triton blowing an Atlantic trumpet triton, which may have inspired the common name. Trumpet tritons feed on sea cucumbers and urchins. The "hairy" tritons are named for the coat of frilly periostracum that covers the living snail. Hairy tritons feed on bivalves.

Nutmeg and Cantharus Snails

Common nutmeg, max 1.7 in (4.5 cm)

RELATIVES: Nutmegs (family Cancellariidae) and cantharus snails (family Buccinidae) are distantly related to auger shells.

IDENTIFYING FEATURES:

Common nutmegs *(Cancellaria reticulata)* are egg-shaped shells with a crosshatched texture and whorls indented at the sutures. The inner lip of the aperture has two white folds on the columella. Shell colors vary between tan with blurry brown streaks and cream-white.

Tinted canthari *(Pollia tincta)* have a similar shape to nutmegs but without distinct whorl indentations. Their outer lip is toothed and the columella is glossy. Background shell color is cream or bluish-gray. Most have streaks and smudges of brown.

Tinted cantharus, max 1.2 in (3 cm)

Ribbed canthari *(Cantharus multangulus)*, also called false drills, have large ridges that are sharply angled at the whorl shoulders.

HABITAT: These snails live in sand, rubble, and seagrass to moderate depths.

DID YOU KNOW? Common nutmegs feed on soft-bodied animals buried in the sand. Canthari get their name from the cantharus, sacred cup of Bacchus, Roman god of wine. Cantharus snails prey on worms, barnacles, and other attached invertebrates.

Ribbed cantharus, max 1.2 in (3 cm)

23

Common American auger, max 2.4 in (6 cm)

Shiny Atlantic auger, max 1.5 in (3.8 cm)

Sallé's auger, max 1.5 in (3.8 cm)

Sanibel turret, max 1 in (2.5 cm)

Augers

Common American   Shiny Atlantic   Sanibel
and Sallé's Augers   Auger   Turret Snail

RELATIVES: Augers (family Terebridae) are distantly related to turrets (family Turridae).

IDENTIFYING FEATURES: Augers are sharp glossy cones with smooth ribs and short, distinct siphon canals.

Common American augers *(Terebra dislocata)* are gray or orange-white with beaded spiral bands between whorls.

Shiny Atlantic augers *(Hastula hastata)* are shiny white with wide orange bands wrapping the bottoms of lower whorls. Their shape is more of a narrow bullet than a sharp cone.

Sallé's augers *(Hastula cinera)* are purple-gray with darker banded ribs below each whorl suture.

Sanibel turrets *(Zonulispira crocata)* have nine ribs crossed by dashlike cords.

HABITAT: Sandy shallows.

DID YOU KNOW? Augers feed on worms. Sallé's auger is an active hunter. It has a long stride and quick pace, nearly one "footstep" per second, and lunges when it finds a worm above the sand. Like other augers, Sallé's subdues its prey with a stab from a venomous, radular tooth. This auger's summer mating swarms are in the style portrayed by Burt Lancaster and Deborah Kerr in *From Here to Eternity,* with embracing pairs rolling in the swash zone.

Mudsnails

Threeline and
Eastern Mudsnails

Bruised
Nassa Snail

White
Nassa Snail

RELATIVES: Mudsnails and nassas are in the family Nassariidae.

IDENTIFYING FEATURES: These are small oval snails with conical spires.

Threeline mudsnails *(Ilyanassa trivittata)* are yellowish-gray with shouldered (stepped) whorls and a basketlike texture.

Eastern mudsnails *(Ilyanassa obsoleta)* are solid brown with smooth, slanting, axial ribs. Their apex is typically worn.

Bruised nassas *(Nassarius vibex)* are light gray to dark with strong rounded axial ribs and a pointed spire. Their inner aperture lip is thickened by a wide glossy callus, which in darker shells bears a purple bruise.

White nassas *(Nassarius albus)* are white with brown on their distinct whorl shoulders.

HABITAT: Mudsnails live in muddy sand at the low-tide line. Bruised nassas are in shallow seagrass and white nassas live on sandy bottom to moderate depths.

DID YOU KNOW? Mudsnails and nassas eat algae, invertebrate eggs, carrion, and other easily outrun items. In Latin, *nassa* means "wicker basket".

Threeline mudsnail, max 7/8 in (22 mm)

Eastern mudsnail, max 1.2 in (3 cm)

Bruised nassa, max 3/4 in (20 mm)

White nassa, max 1/2 in (13 mm)

Tulip Snails

RELATIVES: Tulip snails (family Fasciolariidae) are related to spindle shells.

IDENTIFYING FEATURES: Tulip snails are pointed spindles with rounded curves and a stemlike siphon canal.

Banded tulips *(Fasciolaria lilium)* are cream to light gray with orange or gray splotches and distinctly fine spiral lines of reddish brown. Their whorls are smooth.

True tulips *(Fasciolaria tulipa)* are similar to banded tulips but have darker brown (or orange) splotches and interrupted, closer-set spiral lines. Their whorls also differ in having fine ridges below each suture.

HABITAT: Banded and true tulips live on sand in water less than 100 ft (30 m).

DID YOU KNOW? True tulips prey on banded tulips, as well as on pear whelks and other mid-size gastropods. Tulip snails crawl into shallow waters during the winter to attach their clustered **egg capsules**, which look like tiny bouquets. The young, miniature tulip snails emerge from holes at the flat end of each frilly capsule. The capsules are formed of a tough, fingernail-like protein. If they rattle, the capsules are likely to contain tiny tulip shells. Several occupy each capsule.

Banded tulip, max 4 in (10 cm)

True tulip, max 5 in (13 cm)

Tulip snail egg capsules on a penshell

Spindle Shells

Florida Horse Conch

Chestnut Latirus

RELATIVES: Spindle shells share the family Fasciolariidae with tulip snails.

IDENTIFYING FEATURES: Both of these snails are thick-shelled with knobbed whorls that form a pointed spire about half the total shell-length.

Florida horse conchs *(Triplofusus giganteus)* are unmistakably large as adults. They have a whitish spire and are often covered with brown, flaky periostracum. Beach-worn adult shells are white with a glossy tan interior. Living horse conchs have an orange-red body and a thick operculum (trap door).

Chestnut latiri *(Leucozonia nassa)* are dark golden-brown with lighter spiral cords beneath the knobs of its body whorl.

A **juvenile horse conch** resembles a **chestnut latirus** but is a lighter, more uniform, peach-gold, and has a much longer siphon canal.

HABITAT: Both live in waters as shallow as the low-tide line. Chestnut latiri prefer reefs. Horse conchs prefer sand.

DID YOU KNOW? As the largest snail in North America, horse conchs are able to prey on big gastropods like whelks. This impressive mollusk is Florida's state shell. Horse conch egg masses comprise dozens of flattened bugles clustered in a twisted clump. Chestnut latiri feed on barnacles and worms.

A horse conch's orange body and operculum (A)

Horse conch, max 19 in (48 cm), and egg masses

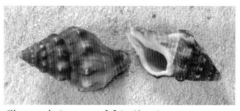
Chestnut latirus, max 2.2 in (6 cm)

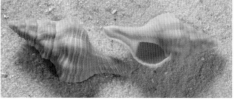
Juvenile horse conch

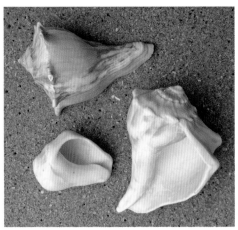

Knobbed whelk, max 9 in (23 cm), new and worn shells

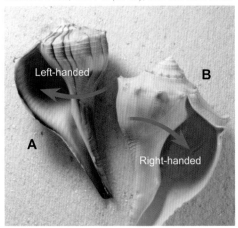

Lightning whelk (A), knobbed whelk (B)

Lightning whelk, max 16 in (40 cm), and egg masses

Whelks *(Knobbed and Lightning)*

Knobbed Whelk   Lightning Whelk

RELATIVES: These snails share the family Melongenidae with pear, channeled, and crown conch whelks.

IDENTIFYING FEATURES: Both species have large body whorls with distinct shoulders and a wide aperture tapering into their long siphon canal.

Knobbed whelks *(Busycon carica)* are light gray to gray-brown and have heavy shells with several triangular knobs on the shoulder of the body whorl. Beach-worn shells often remain intact although every edge is smoothed by surf-sanding.

Lighting whelks *(Busycon sinistrum)* are cream to gray with younger shells showing brown, lightning-bolt, axial streaks. They have a dozen or more petite knobs on their body whorl, which spirals to their left (their spire is rearward). Their left-handed aperture separates them from most other marine snails.

HABITAT: Both live in sandy shallows.

DID YOU KNOW? Knobbed whelks are common on beaches far from where they live. Worn shells from south Florida have probably tumbled south with the longshore current for well over 100 miles (160 km). Both whelks produce egg masses containing dozens of discs attached by a common string. The egg-discs strung together by lightning whelks have edge projections. Those from knobbed whelks are more angled.

Whelks
(Pear, Channeled, and Crown Conch)

Pear Whelk and
Crown Conch

Channeled Whelk

Pear whelk, max 5.5 in (14 cm)

RELATIVES: These snails share the family Melongenidae with other whelks.

IDENTIFYING FEATURES: All have a shouldered body whorl and a wide aperture tapering into their siphon canal.

Pear (fig) whelks *(Busycotypus spiratus)* are cream with brown, wavy, axial streaks. A channel along the suture disappears in earlier whorls.

Channeled whelks *(Busycotypus canaliculatus)* are gray to tan and have a deep channel along their body-whorl suture that goes well into their spire. Their angled whorl shoulders are edged by a lumpy spiral ridge.

Crown conchs *(Melongena corona)* have spiral bands of blue-gray or brown. Their shell is sculptured by axial ribs along the whorl shoulders that commonly project as hollow spikes, but may be only tiny nubs. Spiky shells also have points at their base. Juvenile crown conchs are least spiky.

HABITAT: Channeled whelks live on intertidal sands and nearby shallows. Pear whelks and crown conchs live on muddy sand in shallow, quiet bay waters.

DID YOU KNOW? These snails scavenge and are often captured in baited traps. Knobbed and channeled whelks are the local scungilli served in Italian restaurants.

Channeled whelk, max 7.5 in (19 cm)

Crown conch, max 5 in (13 cm)

29

Glossy dove snail, max 5/8 in (16 mm)

Dove Snails

Glossy and Common   Rusty and Sparse   Greedy
Dove Snails   Dove Snails   Dove Snail

RELATIVES: Dove snails are in the family Columbellidae.

IDENTIFYING FEATURES: Dove snails have tubby shells the size of a pencil eraser with pointed spires, short siphon canals, and toothed aperture lips.

Glossy dove snails *(Nitidella nitida)*, are smooth and splotched with light on brown or brown on light.

Common dove snail, max 7/8 in (22 mm)

Common (West Indian) dove snails *(Columbella mercatoria)* have a triangular body whorl wrapped in rounded spiral cords. Their cords are dashed with white and orange-brown.

Rusty dove snails *(Columbella rusticoides)* have rounded whorls and an outer aperture lip that is tinted and thickened in the center.

Rusty dove snail, max 1/2 in (13 mm)

Greedy dove snails *(Costoanachis avara)* have 12 ribs on their body whorl highlighted by white splotches.

Sparse dove snails *(Costoanachis sparsa)* are orange and white and have 20 or more ribs with crossing cords.

Greedy dove snail, max 3/4 in (20 mm)

HABITAT: Dove snails live in seagrass or rubble out to moderate depths.

DID YOU KNOW? Common and rusty dove snails graze on algae. Greedy and sparse dove snails are carnivores or scavengers. Dove snail eggs are laid within single capsules, and young emerge either swimming or crawling.

Sparse dove snail, max 1/2 in (13 mm)

Murex Shells
(Giant Eastern, Apple, and Lace)

Giant Eastern Murex   Apple and
                                  Lace Murices

Giant eastern murex, max 7 in (18 cm)

RELATIVES: Murices share the family Muricidae with drills and rocksnails.

IDENTIFYING FEATURES: All have highly sculptured shells with round apertures and tubular siphon canals.

Giant eastern murices *(Murex fulvescens)* are turnip-shaped with a body whorl sculptured by about 8 axial ridges (varices), each bearing pronounced hollow spikes. Beached shells are white to gray and may have only worn knobs instead of spikes.

Apple murices *(Chicoreus pomum)* are cream with brown bands and have 3 lumpy varices per whorl. Their inner aperture lip has a thin, flared margin and a dark blotch opposite the siphon canal.

Apple murex, max 4.5 in (12 cm)

Lace murices *(Chicoreus dilectus)* are similar to apple murices but differ in having varices lined with hollow, scoop-shaped spines, and in having a simple, circular, unmarked aperture.

HABITAT: Giant eastern murices live on sand to about 325 ft (100 m) deep. Lace murices live in shallow rubble. Apple murices inhabit intertidal sands.

DID YOU KNOW? These murices are predators of bivalves, which are devoured through holes rasped in their shells. Apple and giant eastern murices prefer oysters.

Lace murex, max 3.2 in (8 cm)

31

Pitted murex, max 1 in (2.5 cm)

Pitted Murex and Drills
(Mauve-mouth, Sharp-ribbed, and Thick-lipped Drills)

Pitted Murex   Mauve-mouth and   Thick-lipped
       Sharp-ribbed Drills   Drill

RELATIVES: Murices and drills are in the murex family, Muricidae.

IDENTIFYING FEATURES: All have sculptured shells with pointed spires.

Pitted murices *(Favartia cellulosa)* are dull white, have 5–7 lumpy varices per whorl, and have a narrow, upturned siphon canal.

Mauve-mouth drills *(Calotrophon ostrearum)* are grayish-white with 7–10 axial ribs and two spiral cords on the lower half of each spire-whorl. The inside aperture is pinkish-purple.

Mauve-mouth drill, max 1.2 in (3 cm)

Sharp-ribbed drills *(Eupleura sulcidentata)* are similar to thick-lipped drills but have smoother whorls and sharper lips.

Thick-lipped drills *(Eupleura caudata)* are pinkish with a long, thin siphon canal, an oval aperture, and a thick, toothed outer lip opposite an equally thick ridge (varix) on the body whorl.

Sharp-ribbed drill, max 7/8 in (22 mm)

HABITAT: Pitted murices and drills live near oyster beds in shallow bays.

DID YOU KNOW? Drills pierce oysters by secreting shell-softening acids and rasping with their toothy radula. The resulting hole is round and tapering to a small pinpoint, just wide enough for the drill to insert digestive enzymes and withdraw oyster soup.

Thick-lipped drill, max 1.5 in (4 cm)

Gulf Oyster Drill and Rocksnail

Gulf Oyster Drill and
Florida Rocksnail

Gulf oyster drill, max 1 in (2.5 cm)

RELATIVES: Drills and rocksnails are in the murex family, Muricidae.

IDENTIFYING FEATURES:

Gulf oyster drills *(Urosalpinx perrugata)* are yellowish with 6–9 large, rounded, axial ribs around each whorl, crossed by numerous spiral cords.

Florida rocksnails *(Stramonita haemastoma)* have sculptured shells and wide apertures with a toothed outer lip. They are whitish to grayish and frequently show red-brown spots. Their shells are highly variable, but all have spiral cords and axial ridges that are most prominent at the shoulders, which may have knobs. Knob size ranges from none to large, aperture color ranges white to orange, and the outer lip interior may have tiny brown lines or fine white ribs. Some of these forms overlap in distribution and may represent as many as three species in Florida.

Florida rocksnail, max 3 in (8 cm)

HABITAT: Florida rocksnails and Gulf oyster drills live in rocky intertidal areas and oyster bars. Rocksnails are common on jetties.

DID YOU KNOW? Rocksnails feed on bivalves, gastropods, and barnacles. Rocksnail eggs are contained in tan or purple-stained, vase-shaped capsules that are attached to rocks during communal gatherings of snails.

A living Florida rocksnail on a jetty boulder

33

Lettered olive, max 2.7 in (7 cm)

Lettered olives show color variation with wear

Variable dwarf olive, max 5/8 in (16 mm)

Olive Shells

RELATIVES: Olive shells are in the family Olividae, distantly related to vase, volute, marginella, and cone shells.

IDENTIFYING FEATURES: All have cylindrical, glossy shells with narrow, elongate apertures.

Lettered olives *(Oliva sayana)* have a thick shell with a small pointed spire about 1/9 of the total shell length. Unfaded shells are covered with overlapping, slightly blurry, brown zigzags that are darkest just below the suture and as part of two, broad, spiral bands. The similar netted olive, *Oliva reticularis* (southeast only), has no bands and a larger spire of about 1/7 of the total shell length.

Variable dwarf olives *(Olivella mutica)* are gray to brown, variably marked, and have a spire nearly half their shell length. Their aperture is triangular with an inner, ridged, parietal callus that extends beyond the aperture up to the next suture. Several other dwarf olive species occur in Florida.

HABITAT: Lettered and variable dwarf olives live within sand in waters as shallow as the low-tide mark.

DID YOU KNOW? An olive's glossy shell is covered in life by its body mantle and large foot. Their strong foot allows them to burrow easily through sand. Lettered olives prey on coquina clams in the surf zone, and both species scavenge when opportunities arise.

Junonia and Vase Shell

Junonia   Vase Shell

Junonia, max 3.5 in (9 cm)

RELATIVES: Junonias are volutes, family Volutidae, and are only distantly related to vase shells, which are in the family Turbinellidae.

IDENTIFYING FEATURES:

Junonias *(Scaphella junonia)* have unmistakably flamboyant shells with squared, chestnut spots on a background of pinkish-ivory. They have a wide aperture and rounded whorls that spin into a distinct spire with a rounded apex.

Caribbean vase shells *(Vasum muricatum)* have dense, cone-shaped shells with short, pointed spires, spiral cords, and numerous blunt knobs at the whorl shoulders. They are dull white with occasional purple tinges.

Beach-worn junonia shells

HABITAT: Junonias live on offshore reefs and Caribbean vases live on shallow reefs.

DID YOU KNOW? Junonia shells are a rare find but they inspire a common quest. Because Florida is the hotspot for this rare and spectacular shell, junonias receive much attention, especially from visitors to Sanibel where, island-wide, roughly one junonia per day is found. Like their shell, the soft parts of the living junonia are covered with spots. Junonias and vases are predators of bivalves.

Caribbean vase shell, max 3 in (8 cm)

35

Florida cone, max 2 in (5 cm)

Cone Shells

Florida and Alphabet Cones *Jasper Cone*

RELATIVES: Cone shells are in the family Conidae, distantly related to volutes, olives, vases, and marginellas.

IDENTIFYING FEATURES: Cone shells are cone-shaped, with long, thin apertures.

Florida cones *(Conus anabathrum)* have a medium-high spire and a body whorl with orange or yellow streaks and splotches on a lighter background.

Jasper cones *(Conus jaspideus)* have a high spire and distinct spiral cords that bear dashes of white and orange or brown.

Alphabet cones *(Conus spurius atlanticus)* have a low, concave spire and spiral rows of orange dots, dashes, and checks.

HABITAT: These cones live in shallow to moderately deep sand and seagrass.

Jasper cone, max 1 in (2.5 cm)

DID YOU KNOW? Cone shells have radular teeth that look and function like harpoons. They use this needle-like weapon to inject venom into the animals they eat, and into would-be attackers. Deadly neurotoxin venoms from fish-eating Indo-pacific cones are used in medicine for treating stroke, heart disease, and chronic pain. Florida's tamer worm-eating species have much milder venom. There are about 400 species of cone shells, 16 of which are found in Florida.

Alphabet cone, max 3 in (7.5 cm)

Melampus Snail and Marginellas

Coffee Malampus, White-spot, and Orange Marginellas   Atlantic Marginella

RELATIVES: Melampus (family Marginellidae) and marginella shells (family Ellobiidae) are distantly related.

IDENTIFYING FEATURES:

Coffee melampi *(Melampus coffea)* have an egg-shaped shell with smooth, broadly conical spires. Their thin outer aperture lip has numerous fine ridges inside. Colors range from pale gray to brown with thin spiral bands.

Atlantic marginellas *(Prunum apicinum)* have a glossy, egg-shaped shell with a low spire and a thick, smooth, outer aperture lip margin extending up past the preceding whorl. Colors range from gray to tan.

White-spot marginellas *(Prunum guttatum)* are pale to brown with numerous light spots.

Orange marginellas *(Prunum carneum)* are orange with a light mid-whorl band.

HABITAT: Coffee melampus snails live at the high-tide line near mangroves. Marginellas live in shallow sand and seagrass areas.

DID YOU KNOW? Several other melampus and marginella species occur in Florida. Melampus snails have a lung instead of gills and spend most of their time out of water. Marginellas get their name from their wide margin (aperture lip). They feed as mostly as scavengers.

Coffee melampus, max 3/4 in (19 mm)

Atlantic marginella, max 1/2 in (13 mm)

White-spot marginella, max 1 in (2.5 cm)

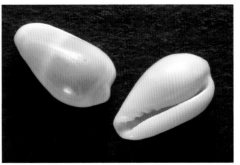

Orange marginella, max 3/4 in (2 cm)

37

Striate bubble shell, max 1 in (2.5 cm)

Gray peanut snail, max 1 in (2.5 cm)

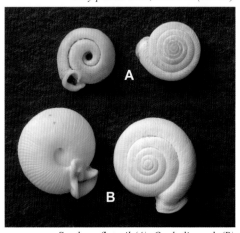

Southern flatcoil (A), Ocala liptooth (B)

Bubble Shell and Land Snails

| Striate Bubble Shell and Southern Flatcoil | Gray Peanut Snail | Ocala Liptooth |

RELATIVES: Bubble shells (family Bullidae) and land snails (order Stylommatophora) are distantly related gastropods. Peanut snails are in the family Ceriidae; liptooths and flatcoils are within the family Polygyridae.

IDENTIFYING FEATURES:

Striate bubbles *(Bulla striata)* have a fragile, smooth, mottled-brown, egg-shaped shell with a sunken apex and an aperture longer than the body whorl.

Gray peanut snails *(Cerion incanum)* are land snails with a lightweight, bullet-shaped shell.

Southern flatcoils *(Polygyra cereolus)* are land snails that have a deep umbilicus and a smooth aperture bearing a single tooth. The related **Ocala liptooth** land snail *(Daedalochila auriculata)* has an aperture like a wrinkled kiss. Each reaches 3/5 in (1.5 mm).

HABITAT: Striate bubbles live in shallow seagrass. The three land snails live on terrestrial plants, including those on the dune.

DID YOU KNOW? Peanut snail shells reach the beach from land thanks to the travels of land hermit crabs. Southern flatcoils are probably the most abundant land snail in Florida. Although these snails are not "seashells," they are occasionally abundant on the upper beach.

Straight-needle Pteropod and Ram's Horn Squid

Straight-needle Pteropod   Ram's Horn Squid

Straight-needle pteropod, max 3/8 in (1 cm)

RELATIVES: Pteropods are gastropods in the order Thecosomata. The ram's horn squid (family Spirulidae) is a cephalopod, most closely related to cuttlefishes, octopods, and nautili.

IDENTIFYING FEATURES:

Straight-needle pteropods *(Creseis acicula)* have glassy, needle-shaped shells that occasionally beach in massive numbers. This tiny, shelled sea slug has paired winglike flaps for swimming.

Ram's horn squid *(Spirula spirula)* are beached as white, chambered coils. The coil lies within the posterior end of the squid opposite its two large eyes and ten tentacles. The coil takes up almost half the squid, minus outstretched tentacles.

Ram's horn squid shell, max 1 in (2.5 cm)

HABITAT: Pteropods inhabit the open ocean surface. Ram's horn squid live in the deep open ocean.

DID YOU KNOW? Pteropods feed by trapping plankton in a mucous web. Ram's horn squid use their buoyant, chambered coil to suspend themselves head-down in the water column. For protection, the squid can pucker up by withdrawing its head and tentacles into its body. They range worldwide.

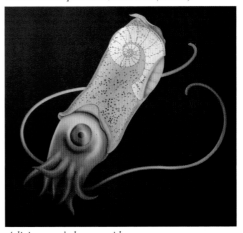

A living ram's horn squid

39

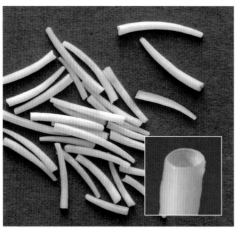

Dentalium *tuskshells, max 1.4 in (3.5 cm)*

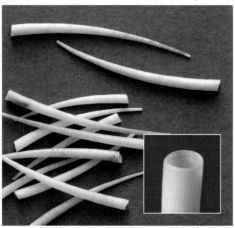

Ivory *tuskshells, max 2 in (5 cm)*

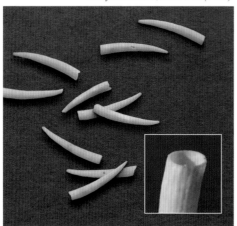

Antalis *tuskshells, max 1.1 in (2.8 cm)*

Tuskshells

RELATIVES: Tuskshells are scaphopods, class Scaphopoda. Although vastly different in shape, tuskshells are thought to be related to bivalves. Both have a reduced head and a retractile foot used for burrowing. These tuskshells are in the family Dentaliidae.

IDENTIFYING FEATURES: Most tuskshells are pale, delicate, curved, tapered tubes open at each end. The foot and mouth of the living tuskshell were formerly located at the wide end. The smaller rear opening passed respiratory currents.

Dentalium (spp.) tuskshells are commonly hexagonal in cross-section.

Ivory tuskshells *(Graptacme eborea)* are off-white, gray, or pale peach with a relatively sharp tip. They are smooth and round in cross-section.

Antalis (spp.) tuskshells are commonly ribbed and round in cross-section.

HABITAT: Some tuskshell species live in shallow bay sediments and others live in deep offshore seabottom.

DID YOU KNOW? Tuskshells live with their wide (anterior) end in the seabottom where they use their oral tentacles to feed on forams (protozoa). Their pointed rear is often their only exposed feature.

Arks
(Blood, Transverse, and Ponderous)

Blood ark, max 3 in (7.5 cm)

RELATIVES: Ark shells are allied within the family Arcidae.

IDENTIFYING FEATURES: All ark shells are thick to very thick with forward umbones and distinct ribs.

Blood arks *(Anadara ovalis)* have a very thick, oval shell with an arched hinge-line bearing about 7 teeth in front of the umbo and about 30 teeth behind. The most rearward hinge teeth are largest and angled backward.

Transverse arks *(Anadara transversa)* have an elongate oval shell with a relatively straight hinge-line bearing mostly vertical teeth below a thin ligament scar the length of the hinge.

Ponderous arks *(Noetia ponderosa)* have a very thick, triangular shell with flat, divided ribs and an arched hinge-line below a broad, moustachelike ligament area.

Transverse ark, max 1.4 in (3.6 cm)

HABITAT: These arks live in nearshore sands as shallow as the low-tide line.

DID YOU KNOW? These ark shells are coated with a brown, fuzzy periostracum in life, worn white after beaching, or may be stained gray, rust, or black. Robust shells make blood and ponderous arks among the most common whole shells on high-energy beaches where waves pulverize most other mollusks. The blood ark gets its name from its uncommon, hemoglobin-red blood.

Ponderous ark, max 2.8 in (7 cm)

41

Incongruous ark, max 3 in (7.5 cm)

Eared ark, max 3 in (7.5 cm)

Cut-ribbed ark, max 4.5 in (11.5 cm)

White miniature ark, max 7/8 in. (22 mm)

Arks *(Incongruous, Eared, Cut-ribbed, and White Miniature)*

Incongruous Ark

Eared, Cut-ribbed, and White Miniature Arks

IDENTIFYING FEATURES:

Incongruous arks *(Anadara brasiliana)* are thin-shelled for an ark. Their strong radial ribs have distinct dash-shaped beads.

Eared arks *(Anadara notabilis)* have shells with rounded ribs and a long, straight hinge-line that angles sharply with the front and rear edges. The upper rear of unworn shells has a flattened, pointed "ear."

Cut-ribbed arks *(Anadara floridana)* are similar to eared arks but have flattened ribs. Each of the forward ribs (toward umbo) has a deep central "cut."

White miniature arks *(Barbatia domingensis)* have rear-pointed shells with heavy growth lines that give it a scaly look.

HABITAT: Incongruous and eared arks live within shallow sandy bottom. Cut-ribbed arks live in offshore sands but are closer to shore in south Florida. White miniature arks grow under shallow-water rubble.

DID YOU KNOW? Young eared arks attach themselves to small bits of rubble using a threadlike byssus. As they grow they become unattached burrowers. White miniature arks remain attached for life.

Arks *(Mossy, Turkey Wing, and Red-brown)*

Mossy and Turkey Wing Arks   Red-brown Ark

Mossy ark, max 2.5 in (6.3 cm)

IDENTIFYING FEATURES: These arks have elongate, rectangular shells with long, straight, hinge-lines below a broad, triangular ligament area.

Mossy arks *(Arca imbricata)* have shells with beaded ribs and are mostly chestnut brown.

Turkey wings (zebra arks) *(Arca zebra)* have shells with rough ribs, but less beaded than in the mossy ark. Turkey wing shells are striped by nested, red-brown V's or W's at the umbo that turn into oblique lines or zig-zags rearward.

Red-brown arks *(Barbatia cancellaria)* are finely beaded and brown except for a light streak from the umbo down.

Turkey wing, max 3.6 in (9 cm)

HABITAT: These arks live attached by byssal threads to shallow-water rubble.

DID YOU KNOW? The arched shell-gap opposite the umbo in these arks marks the opening where byssal threads anchor them to the bottom. In the Caribbean, thousands of tons of turkey wing arks are harvested each year to be canned and eaten. These filter-feeding bivalves are targets in one of the most important fisheries in Venezuela. Both mossy and turkey wing arks are ecologically important as plankton feeders, prey, and substrate for benthic animals.

Red-brown ark, max 1.3 in (3.3 cm)

Giant bittersweet clam, max 4 in (10 cm)

Spectral bittersweet clam, max 1.5 in (3.8 cm)

Comb bittersweet clam, max 1.2 in (3 cm)

Bittersweet Clams

RELATIVES: Bittersweets are in the family Glycymerididae, and may be distantly related to the arks.

IDENTIFYING FEATURES: Bittersweet clams have heavy, rounded shells with thick, arching hinge-lines bearing several prominent teeth on either side of the umbo.

Giant bittersweets *(Glycymeris americana)* have a nearly circular shell with a small, rounded umbo and roughly 50 flattened ribs. They are glossy cream with concentric, blurry necklaces of tan or rust.

Spectral bittersweets *(Glycymeris spectralis)* have a slightly triangular (or elliptical) shell with a large umbo and 30–40 smooth ribs. They range from white to brown and frequently show varied radiating sunburst streaks of white on chestnut, or vice versa.

Comb bittersweets *(Glycymeris pectinata)* have a nearly circular shell with a small, relatively pointed umbo, and 20–30 raised ribs. They are slightly roughened by growth lines and are grayish-white with brownish spatters.

HABITAT: These bittersweet clams live in sand from shallow moderate depths.

DID YOU KNOW? Bittersweets are indeed named for their taste and are elements of recipes in many eastern Atlantic countries. They live unattached and have light-sensitive eyespots along their mantle.

Mussels

Ribbed Mussel   *Scorched, Horse, and Green Mussels*

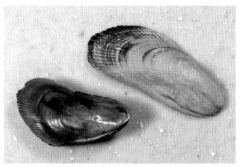

Ribbed mussel, max 5 in (13 cm)

RELATIVES: Mussels (family Mytilidae) are distantly related to penshells.

IDENTIFYING FEATURES: Mussels have thin shells that fan out from their umbones and tend to retain their thin, shiny periostracum.

Ribbed mussels *(Geukensia demissa)* have shells with radiating ribs and no hinge teeth. Shells without the brown periostracum are yellowed gray with occasional purple tinges.

Scorched mussels *(Brachidontes exustus)* have shells with radiating ribs and 2 or 3 hinge teeth under the umbo.

Scorched mussel, max 1.5 in (4 cm)

Horse mussels *(Modiolus spp.)* have inflated shells and an umbo just shy of their upper end. Two similar species occur. **Southern horse mussels** *(M. squamosus)* reach 2.5 in (6.5 cm), have less-inflated umbones, and are whitish or purple after beach wear. **American horse mussels** *(M. americanus)* reach 4 in (10 cm), have bulbous umbones, and are bright red through their golden periostracum.

Southern (A) and American (B) horse mussels

Green mussels *(Perna viridis)* have a smooth green-and-brown exterior.

HABITAT: These mussels grow attached to rocks or pilings in estuarine waters as shallow at the intertidal zone.

DID YOU KNOW? Green mussels are alien invaders from Asia.

Green mussel, max 3.5 in (9 cm)

45

Sawtooth penshell, max 9 in (23 cm)

Half-naked penshell, max 9 in (23 cm)

Penshells *(Sawtooth and Half-naked)*

Sawtooth Penshell   Half-naked Penshell

RELATIVES: Penshells (family Pinnidae) are distantly related to mussels.

IDENTIFYING FEATURES: Penshells have thin, amber-brown, fanlike valves.

Sawtooth penshells *(Atrina serrata)* have about 30 radiating ribs bearing hundreds of short, hollow prickles.

Half-naked penshells *(Atrina seminuda)* have about 15 radiating ribs bearing a few to dozens of long tubular spines. Their posterior (fan end) muscle scar is completely within their pearly (or cloudy) "nacreous" area.

HABITAT: Penshells live in colonies with individuals buried in soft sediment out to 20 ft (6 m).

DID YOU KNOW? Penshells anchor themselves with golden byssal threads, which lead from their pointed (front) end to a small bit of rubble beneath the sand. Like most bivalves, they are filter feeders. Many living penshells have pale, soft-bodied pen shrimp or pea crabs living within their mantle cavity. The crabs and shrimp feed on items filtered from the water by their penshell host.

Penshells *(Stiff and Amber)*

Stiff Penshell   Amber Penshell

Stiff penshell, max 11 in (28 cm), arrow shows scar

RELATIVES: Penshells (family Pinnidae) are distantly related to mussels.

IDENTIFYING FEATURES:

Stiff penshells *(Atrina rigida)* have 15–20 radiating ribs and tubular spines similar to half-naked penshells but are darker, broader, and have their posterior muscle scar outside the shiny nacre.

Amber (flesh) penshells *(Pinna carnea)* are thin, fragile and tinged with yellow or rose. Their shiny nacre inside is separated into two lobes by a narrow groove, and their hinge (dorsal) side much longer than the opposite open (ventral) side.

HABITAT: Penshells live in colonies with individuals buried in soft sediment out to 20 ft (6 m).

DID YOU KNOW? Penshells are harvested for food in Mexico, where the muscles that close the shell are prepared like scallops. The intense commercial harvest has resulted in declining populations. In Japan, penshells are raised in aquaculture. The shells occasionally produce a "pinna pearl," which may be either orange or silvery iridescent.

Amber penshell, max

47

Atlantic wing oyster, max 3.5 in (9 cm)

Atlantic pearl oyster, max 3.5 in (9 cm)

Flat tree-oyster, max 4 in (10 cm)

Oysters

Atlantic Wing and Atlantic Pearl Oysters

Flat Tree-oyster

RELATIVES: Wing and pearl oysters, family Pteriidae, are related to flat tree-oysters, family Isognomonidae.

IDENTIFYING FEATURES: These oysters have straight hinges and thin shells covered with scaly periostracum.

Atlantic wing oysters *(Pteria colymbus)* have valves with a triangular front wing near the umbo and a long rear wing extending past the rest of the shell.

Atlantic pearl oysters *(Pinctada imbricata)* have valves with short triangular front and rear wings and have a scaly, fringelike periostracum.

Flat tree-oysters *(Isognomon alatus)* have both valves equally flat and a straight hinge with 8–12 distinct grooves.

HABITAT: These free-swinging bivalves live attached by their byssal threads. Atlantic wing and Atlantic pearl oysters are attached mostly to offshore soft corals. Flat tree-oysters attach to shallow-water rubble and mangrove prop roots.

DID YOU KNOW? These oysters are most commonly found after storms have washed in the soft corals, mangrove roots, and trap-float lines on which they grow. Southern Caribbean pearl oysters produce pearls and were harvested to depletion by the Spanish in the late 1500s.

Scallops *(Calico and Bay)*

Atlantic Calico Scallop   Bay Scallop

RELATIVES: Scallops are allied within the family Pectinidae.

IDENTIFYING FEATURES: Scallops have round or oval shells with distinct ribs and winglike "ear" projections on either side of the umbo.

Atlantic calico scallop, max 2.7 in (7 cm)

Atlantic calico scallops *(Argopecten gibbus)* have shells with 19–21 rounded ribs. Shell colors vary through white, yellow, orange, red, purple, and gray, generally with splotches of dark on light. Their ears are often worn.

Atlantic bay scallops *(Argopecten irradians)* have shells with 17–18 ribs that are squarish in comparison to calico scallops. Shell color may be white, gray-brown, or orange.

HABITAT: Atlantic calico scallops live on sand bottom at depths to 1300 ft (400 m). Bay scallops live on muddy sands and seagrass in shallow waters.

Atlantic bay scallop, max 4 in (10 cm)

DID YOU KNOW? The palest scallop shells are the right (lower) valves. Due to overharvest and habitat loss, once-plentiful bay scallops are scarce in Florida except for the Big Bend region. A fishery for calico scallops off Cape Canaveral has taken as much as 40 million pounds (18 million kg) per year, but population busts bring about years when few are harvested. Scallops filter-feed on bits of organic stuff, and are eaten by gastropods, squid, octopi, sea stars, crabs, and people.

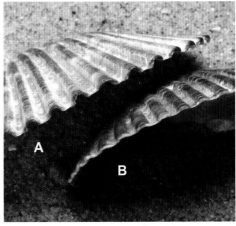

Bay scallop (A) and calico scallop (B)

49

Scallops
(Antillean, Zigzag, and Round-rib)

Antillean Scallop

Zigzag and Round-rib
Scallops

IDENTIFYING FEATURES:

Antillean scallops *(Bractechlamys antillarum)* have shells colored yellow to peach with occasional purplish streaks. They have about 10 broad ribs separated by several fine lines.

Zigzag scallops *(Euvola ziczac)* have a domelike lower valve and an upper valve that looks as if it were melted flat. Their color pattern has distinct zigzags of purple on a background of cream, orange, or light purple.

Round-rib scallops *(Euvola raveneli)* are similar to zigzag scallops but have rounder, separated ribs on the flat, upper valve. The flat valve varies from light gray to purple with rayed streaks.

HABITAT: Antillean scallops live in shallow coraline sands. Round-rib and zigzag scallops live in waters as deep as 330 ft (100 m).

DID YOU KNOW? Round-rib and zigzag scallops lay with their flat upper valve flush with the sandy bottom. If disturbed by a potential predator, they launch from the bottom by clapping their valves to create jets of propulsion. Scallops are the swiftest of the bivalves. Most have numerous light-sensing eyes lining their mantle to detect predators and perhaps even steer their swimming.

Antillean scallop, max 2 in (5 cm)

Zigzag scallop, max 4 in (10 cm) lower valve (A)

Round-rib scallop, upper valve, max 2 in (5 cm)

Scallops *(Scaly and Rough)*

Scaly Scallop	Rough Scallop

Scaly scallop, max 1.6 in (4.1 cm)

IDENTIFYING FEATURES:

Scaly scallops *(Caribachlamys sentis)* have fan-shaped shells with numerous, finely beaded ribs and a front ear about 5 times longer than the rear one. They are most often a mottled orange-red but may also be pale, light orange, purple, or combinations of these colors.

Rough scallops *(Lindapecten muscosus)* have a rounder shape than scaly scallops, less lopsided ear lengths, and fewer (only about 19) ribs that are roughened by tiny spoon-shaped prickles. Beach-worn shells are less prickly. Most rough scallops are solid-colored lemon, peach, or tangerine, but some are mottled with plum.

Rough scallop, max 1.5 in (4 cm)

HABITAT: Scaly scallops live attached beneath rubble in shallow to moderate depths. Rough scallops are free-living on offshore sand banks.

DID YOU KNOW? The eyes lining each mantle of the scaly scallop are as red as its shell. Scallops also have long tentacles for touch and taste that fringe their open shell. Both of these scallops are only moderately common but are conspicuous on the beach due to their sanguine colors.

A rough scallop's rib prickles

51

Lion's-paw, max 6 in (15.2 cm)

Atlantic kittenpaw, max 1.2 in (3 cm)

Atlantic kittenpaw color and shape variation

Scallops *(Lion's-paw and Atlantic Kittenpaw)*

Lion's-paw Scallop

Atlantic Kittenpaw Scallop

IDENTIFYING FEATURES:

Lion's-paws *(Nodipecten nodosus)* have thick, flattened shells with 7–8 large, roughly ridged ribs bearing occasional hollow knuckles. Their outer shell color is commonly orange or brick red, but may range from pale to purple.

Atlantic kittenpaws *(Plicatula gibbosa)* have thick, tough, flattened shells with 6–10 curving, digitlike ribs. They are white to gray except for their tabby-orange ribs marked with numerous, thin, red-brown lines.

HABITAT: Lion's-paws live offshore in sandy rubble at depths up to 160 ft (50 m). Atlantic kittenpaws live attached to rocks in waters from intertidal depth to 300 ft (91 m).

DID YOU KNOW? Lion's-paws are impressive and rare enough to be a quest shell for many beachcombers. Although bits and pieces are moderately common on the Atlantic coast, whole shells are an unusual find. Kittenpaws are common on beaches in part due to their toughness. Left-valve kittenpaws are most common because the right valve often remains attached where the animal lived. The flat attachment site on the right valve retains an impression of the shell or rock on which it grew.

Atlantic Thorny Oyster, Common Jingle Shell, and Crested Oyster

Atlantic Thorny Oyster

Common Jingle Shell and Crested Oyster

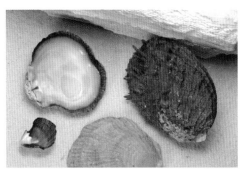
Atlantic thorny oyster, max 5.1 in (13 cm)

RELATIVES: Atlantic thorny oysters (family Spondylidae) are related to jingle shells (family Anomiidae) and crested oysters (family Ostreidae).

IDENTIFYING FEATURES:

Atlantic thorny oysters *(Spondylus* spp.) have thick, circular, lumpy valves with occasional thorns (long in unworn shells). The hinge on the cup-shaped lower valve has two large cardinal teeth separated by a split, and the upper valve hinge has two corresponding sockets. Colors are commonly orange and brick red.

Common jingle shells *(Anomia ephippium)* have round, translucent shells with no clear umbo or hinge. Their colors include silver-gray, white, yellow, and orange. Black shells have been stained by sulfurous sediments.

Crested oysters *(Ostreola equestris)* have oval, lumpy shells. Unworn shells have a rayed, flaky periostracum.

HABITAT: Atlantic thorny oysters attach to rocky reefs and include two species: *S. ictericus* (shallow) and *S. americanus* (deep). Common jingle shells and crested oysters live attached in shallow marine waters to rocks, wood, and other shells.

DID YOU KNOW? Nearly all beached jingle shells are the unattached left valve.

Common jingle shell, max 2 in (5 cm)

Crested oyster, max 2 in (5 cm)

Crested oysters attached to beached flotsam

53

Eastern oyster, max 6 in. (15 cm)

Frond oysters on sea whips, max 2.7 in (7 cm)

Frond oysters on a trap float

Antillean lima, max 1.1 in (2.8 cm)

Oysters and Antillean Lima

Eastern Oyster   Frond Oyster   Antillian Lima

RELATIVES: Oysters are in the family Ostreidae and limas are in the family Limidae.

IDENTIFYING FEATURES:

Eastern oysters *(Crassostrea virginica)* have lumpy shells that vary from oval to clown-shoe shapes. Their inner surface is smooth with a purple muscle scar.

Frond oysters *(Dendostrea frons)* have yellow- or purple-colored oval shells with strong radial ridges ending in interlocking scalloped margins. Those attached by fingerlike shell projections to the branches of soft corals have the most elongate shell shape.

Antillean limas *(Limaria pellucida)* have thin white shells with fine riblets.

HABITAT: These oysters live in shallow waters attached to rocks, debris, or other oysters. Eastern oysters prefer brackish waters. Frond oysters often grow on sea whips. Limas live in crevices.

DID YOU KNOW? Eastern oysters are common near inlets and in seafood raw bars. Frond oysters are beached with storm-tossed corals and trap parts. All are filter feeders and are preyed upon by a variety of animals. Northeast Florida barrier islands have 4000-year-old Timucuan middens piled 50 ft (15 m) high with eastern oyster shells.

Lucines *(Buttercup, Thick Buttercup, and Pennsylvania)*

Buttercup lucine, max 2.5 in (6.4 cm)

RELATIVES: Lucines are allied within the family Lucinidae.

IDENTIFYING FEATURES: Lucines have thick circular shells with forward-pointing umbones above a distinct, heart-shaped impression (the lunule) split by the valve opening.

Buttercup lucines *(Anodontia alba)* have a forward flare that forms a keel protruding more than the umbo. The outer shell is dull white with fine growth lines and the inner shell is butter yellow or cream.

Thick buttercup lucines *(Lucina pectinata)* are similar to buttercup lucines but are thicker shelled, more compressed, and have coarser growth lines. A furrow creases behind the umbo, which protrudes about as much as the keel in front of it. Colors are generally on the pale side of yellow or orange.

Thick buttercup, max 2.7 in (6.7 cm)

Pennsylvania lucines *(Lucina pensylvanica)* have very thick, off-white valves with a deep furrow either side of the umbo. Thin, scaly, growth lines are separated by smooth bands.

HABITAT: All live in sandy shallows. Buttercup lucines and Pennsylvania lucines can live as deep as 300 ft. (90 m).

DID YOU KNOW? Lucines are named for Lucina, an aspect of the Roman goddess Juno, who represented light and childbirth.

Pennsylvania lucine, max 2 in (5 cm)

55

Tiger lucina, max 3.5 in (8.9 cm)

Cross-hatched lucine, max 1 in (2.5 cm)

Cross-hatched lucine detail

Lucines *(Tiger Lucina and Cross-hatched Lucine)*

Tiger Lucina	*Cross-hatched Lucine*

IDENTIFYING FEATURES:

Tiger lucinas *(Codakia orbicularis)* are compressed with thick, yellow-white valves sculptured with fine riblets and concentric growth ridges. Their lunule is a small deep pit beneath a pointed umbo. The insides of the youngest shells are yellow, rimmed with pink.

Cross-hatched lucines *(Divaricella quadrisulcata)* are moderately inflated with relatively thin valves sculptured by numerous, parallel lines that make the shell appear covered with fingerprints. Beached shells are glossy white, chalky, or ivory.

HABITAT: Both of these lucines live in muddy sand in shallow waters. Cross-hatched lucines occur the deepest to about 300 ft (90 m).

DID YOU KNOW? Tiger lucinas are most common on beaches with coraline sands. Lucines filter plankton and detritus from water drawn into a mucus-lined tube that is maintained by their long foot.

Jewelboxes and Cardita

Florida Spiny Jewelbox and Broad-ribbed Cardita

Leafy and Corrugate Jewelboxes

RELATIVES: Jewelboxes (family Chamidae) and carditas (family Carditidae) are allied with clamlike bivalves.

IDENTIFYING FEATURES:

Florida spiny jewelboxes *(Arcinella cornuta)* are shaped like tubby commas bearing about 8 radiating ridges with hollow spines (or knobs, if beach-worn). They are white with a pinkish interior.

Leafy jewelboxes *(Chama macerophylla)* have thick, oval shells covered in numerous scaly ridges. Beach-worn shells are lumpy, but new shells may have long, hollow scales. They are generally yellow or chalky, but are often orange or lavender.

Corrugate jewelboxes *(Chama congregata)* have a corrugated exterior and fine ridges within the inner valve margins, and are reddish outside, purplish inside.

Broad-ribbed carditas *(Carditamera floridana)* have very thick valves with about 15 strong, beaded ribs. They are most often white with bands of chestnut.

HABITAT: Jewelboxes live cemented to reefs and debris to moderate depths. Florida spiny jewelboxes detach when young to grow free within sandy rubble. Carditas live in shallow sands.

DID YOU KNOW? Spines and scales help bivalves avoid being drilled by gastropod predators.

Florida spiny jewelbox, max 2.5 in (6.3 cm)

Leafy jewelbox, max 3.1 in (8 cm)

Corrugate jewelbox, max 1 in (2.5 cm)

Broad-ribbed cardita, max 2.5 in (6.3 cm)

57

Atlantic strawberry-cockle, max 2 in (5.1 cm)

Atlantic giant cockle, max 5.2 in (13.2 cm)

Spiny papercockle, max 1.8 in (4.5 cm)

Cockles

Atlantic Strawberry-cockle   Atlantic Giant Cockle
and Spiny Papercockle

RELATIVES: Cockles are allied within the family Cardiidae.

IDENTIFYING FEATURES: Cockle shells are oval, inflated, and have a large umbo with one central tooth and socket.

Atlantic strawberry-cockles *(Americardia media)* are cream with red-brown specks and have numerous flattened ribs that feel like sandpaper. An angled ridge runs from the umbo across the longest part of the shell.

Atlantic giant cockles *(Dinocardium robustum)* are cream with brown or tan in segments along their shell ribs, which are rounded and bumpy on one side (the front) flattened and smooth on the other (the rear).

Spiny papercockles *(Papyridea soleniformis)* are compressed for a cockle with rear ribs ending in protruding spines. They are mottled with pale pink, purple, orange, or red-brown.

HABITAT: These cockles live in sandy shallows off beaches. Atlantic giant cockles may occur as deep as 100 ft (30 m).

DID YOU KNOW? Giant cockles are also called heart cockles. Shells of this species found on Gulf beaches have more pronounced color patterns than Atlantic shells and are often separated as a subspecies called Van Hyning's cockle *(Dinocardium robustum vanhyningi)*.

Pricklycockles

Florida and
Yellow Pricklycockles

Even Pricklycockle

Florida pricklycockle, max 2.7 in (6.9 cm)

RELATIVES: Pricklycockles are allied with other cockles in the family Cardiidae.

IDENTIFYING FEATURES: Prickly-cockles have inflated valves with a large umbo and a lone tooth and socket. Their ribs bear sharp, cuplike scales, which are reduced to bumps when beach-worn.

Florida pricklycockles *(Trachycardium egmontianum)* have about 30 pro-nounced ribs covered by strong scales (in unworn shells) and ending in a hind margin that is deeply serrated. Their external color is cream with tan or pur-ple-brown splotches. Their valves inside are salmon and/or purple.

Even pricklycockles *(T. isocardia)* are similar to Florida pricklycockles but are heavier, more elongate, and have more ribs (about 35).

Even pricklycockle, max 3 in (7.5 cm)

Yellow pricklycockles *(T. muricatum)* have about 35 ribs with small scales. They have tinges of yellow inside and out and may tend toward peach with occasional red-brown streaks.

HABITAT: These pricklycockles live in sandy shallows near beaches and out to about 100 ft (30 m).

DID YOU KNOW? Cockles live just beneath the sand surface and are often washed ashore after storms. Their prick-les may help anchor them in place, or they may deter gastropod predators, or both.

Yellow pricklycockle, max 2.5 in (6.4 cm)

59

Common egg cockle, max 3 in (7.6 cm)

Painted egg cockle, max 1 in (2.5 cm)

Morton's egg cockle, max 1 in (2.5 cm)

Egg Cockles

Common Egg Cockle   Painted and Morton's
                           Egg Cockles

RELATIVES: Egg cockles are allied with other cockles in the family Cardiidae.

IDENTIFYING FEATURES: Egg cockles are smooth with only faint riblets.

Common egg cockles *(Laevicardium laevigatum)* have valves with an oblique oval shape and ridges along the inner margin. They are glossy white or yellow with occasional rosy tinges. Older beached shells are white and less glossy.

Painted egg cockles *(Laevicardium pictum)* are compressed for a cockle and have a triangular-like shape. They are cream with blurry zigzags and spatters of brown or yellow-orange.

Morton's egg cockles *(Laevicardium mortoni)* are almost evenly rounded with a central umbo and are colored by relatively distinct rows of brown, purple, or orange zigzags.

HABITAT: Common and painted egg cockles occur in sandy areas out to moderate depths. Morton's egg cockles prefer shallow inlet areas and lagoons.

DID YOU KNOW? These cockles can literally leap about using their muscular foot. The response does not always allow Morton's egg cockle to escape ducks, who have this bivalve on their favorite-foods list.

Fragile Atlantic Mactra Clam, Southern Surfclam, and Dwarf Surfclam

Fragile Atlantic mactra clam, max 4 in (10.2 cm)

RELATIVES: Mactra and surfclams are allied within the family Mactridae.

IDENTIFYING FEATURES: These clams have a spoon-shaped pit behind the central hinge teeth.

Fragile Atlantic mactras *(Mactra fragilis)* are relatively thin-shelled with a forward umbo. Beached shells are cream with some remaining periostracum behind a ridgeline on the hind end.

Southern surfclams *(Spisula raveneli)* have strong shells with a central umbo and fine growth lines. They range from white to dirty cream with rusty tones.

Dwarf surfclams *(Mulinia lateralis)* have an umbo forward of center and a tapered hind end. Colors may be white, cream, gray, or purple-gray, with highlighted growth bands.

HABITAT: All live in sand from just off the beach out to moderate depths (165 ft or 50 m). Dwarf surfclams are also common in shallow lagoons.

DID YOU KNOW? Some surfclam species are commercially harvested for food in the southeast US. The abundant dwarf surfclam feeds a host of estuarine animals including ducks, and a hefty shell-crunching fish called the black drum *(Pogonias cromis)*.

Southern surfclam, max 5.1 in (13 cm)

Dwarf surfclam, max 3/4 in (20 mm)

Channeled duckclam, max 3.2 in (8.1 cm)

Smooth duckclam, max 3 in (7.6 cm)

Atlantic rangia clam, max 2.7 in (7 cm)

Duckclams and Atlantic Rangia

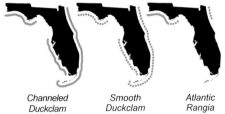

| Channeled Duckclam | Smooth Duckclam | Atlantic Rangia |

RELATIVES: These clams are allied with surfclams in the family Mactridae.

IDENTIFYING FEATURES: These clams all have a distinct spoon-shaped pit behind the central hinge teeth.

Channeled duckclams *(Raeta plicatella)* have white, thin, ear-shaped shells with strong concentric growth ridges.

Smooth duckclams *(Anatina anatina)* have thin, off-white, ear-shaped shells with relatively smooth growth lines. Their highly flared hind end has a distinct ridge leading from the umbo.

Atlantic rangia clams *(Rangia cuneata)* have thick shells with inflated, forward-pointing umbones. Their front hinge tooth is large and rectangular and their rear tooth is a long, flat ridge. Beach shells are white, yellow, variously stained, and generally worn.

HABITAT: Channeled duck clams live in sand just outside the surf zone. Smooth duck clams live offshore to depths of about 250 ft (75 m). Atlantic rangia live in the muddy sands of brackish bays.

DID YOU KNOW? Duckclams gape open at their flared hind end from which their siphons protrude. Atlantic rangia are abundant as fossils and in Indian middens, and are a base material for many Gulf coast roadbeds.

Tellins *(Alternate, Rose Petal, Sunrise, and Candystick)*

Alternate Tellin | Rose Petal and Candystick Tellins | Sunrise Tellin

Alternate tellin, max 2.7 in (6.9 cm)

RELATIVES: Tellins are allied with macomas in the family Tellinidae.

IDENTIFYING FEATURES: Valves in each of these species have a rounded front end and a tapered rear that has a recognizable rightward bend at the rear (outward in the right valve).

Alternate tellins *(Tellina alternata)* are a pearly yellow-white with numerous concentric grooves between flattened concentric ridges. They may have yellow or pink radiating from the umbo.

Rose petal tellin, max 1.3 in (3.3 cm)

Rose petal tellins *(Tellina lineata)* are rosy white to deep pink.

Sunrise tellins *(Tellina radiata)* are elongate with glossy, ivory shells that may be rayed with yellow and/or pink, emanating from a pink-tipped umbo.

Candystick tellins *(Tellina similis)* have thin, pearly shells with pink rays.

HABITAT: Each of these tellins lives in sand off beaches out to moderate depths.

Sunrise tellin, max 4.5 in (11.4 cm)

DID YOU KNOW? Tellins lie beneath the sand on their left valve so that their posterior curves upward. This accommodates their long intake siphon, which draws in surface morsels. Their blade-like form and strong foot allow rapid burrowing should a predator approach. Like the arks, tellin bodies are red from the oxygen-binding pigment, hemoglobin.

Candystick tellin, max 1 in (2.5 cm)

Speckled tellin, max 2.3 in (6 cm)

Favored tellin, max 4 in (10 cm)

Tampa tellin, max 1 in (2.5 cm)

Tellins *(Speckled, Favored, and Tampa)*

Speckled Tellin   Favored Tellin   Tampa Tellin

IDENTIFYING FEATURES: Like other tellins, each of these shells is less rounded in the rear where they bend right (outward in the right valve).

Speckled tellins *(Tellina listeri)* have strong concentric growth lines and two ridges at their posterior (pointed) end where they bend sharply right. Inside they are yellow and outside they are cream with blurry brown zigzags.

Favored tellins *(Tellina fausta)* have large, thick shells with crowded growth lines that are coarse near the outer margin. They are white with an occasional hint of yellow inside and at the umbo.

Tampa tellins *(Tellina tampaensis)* have thin, strong, cream-colored shells that are smooth except for outer growth lines.

HABITAT: Speckled and favored tellins live in sand or seagrass out to moderate depths (100 ft, 30 m). Tampa tellins prefer shallow bays and lagoons.

DID YOU KNOW? The speckled tellin is also known as Lister's tellin for Martin Lister, a British medical doctor who in 1685 published the first detailed book on shells. Favored tellins are favored by octopi, who find them delicious. Tampa tellins are often the dominant mollusk in enclosed hypersaline lagoons. They are one of many shells that only occasionally make it to the beach even though they are highly abundant where they live.

Coquina Clams and Minor Jackknife Clam

Variable Coquina and
Minor Jackknife Clams

Giant
Coquina Clam

Living variable coquina clams, max 1 in (2.5 cm)

RELATIVES: Coquina clams (family Donacidae) are related to tellins. Minor jackknife clams (family Cultellidae) are related to razor clams.

IDENTIFYING FEATURES:

Variable coquina clams *(Donax variabilis)* have glossy, wedge-shaped shells with faint riblets and groove-teeth lining their inner margins. Patterns vary between solids, radial rays, and concentric bands, and may include any color.

Giant coquina clams *(Iphigenia brasiliana)* have smooth, thick shells with an umbo just rear of center. Worn shells are cream with hints of purple inside.

Minor jackknife clams *(Ensis minor)* have fragile shells with a curved straight-razor shape, purplish inside and whitish outside. The similar green jackknife (*Solen viridis*, family Solenidae, north Florida) has a straight upper shell edge.

HABITAT: Variable coquinas live in the swash zone. Giant coquinas live in deeper water out to about 13 ft (4 m).

DID YOU KNOW? Variable coquinas are one of the most abundant and ecologically important mollusks on Florida beaches. Specialized for life in wave-washed sand, they filter-feed on algae and bacteria swept to shore. They are a critical food for shore birds and surf fishes.

Giant coquina clam, max 2.6 in (6.6 cm)

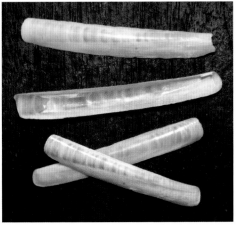

Minor jackknife clam, max 3 in (7.6 cm)

65

Purplish semele, max 1.5 in (3.8 cm)

Cancellate semele, max 3/4 in (1.9 cm)

White Atlantic semele, max 1.5 in (3.8 cm)

Tellin semele, max 1 in (2.5 cm)

Semeles

Purplish and White Atlantic Semeles / Cancellate Semele / Tellin Semele

RELATIVES: Semeles (family Semelidae) are related to tellins and coquinas.

IDENTIFYING FEATURES: Like tellins, a semele's hind end bends right. Their hinges have a diagonal depression angling back from the umbo.

Purplish semeles *(Semele purpurascens)* have smooth oval shells with an umbo toward the rear. They have smudge-streaks of blurry purple, brown, or orange.

Cancellate semeles *(Semele bellastriata)* are cream or gray with concentric ridges and radial riblets front and rear.

White Atlantic semeles *(Semele proficua)* have a central umbo and are cream with occasional nervous purple lines.

Tellin semeles *(Cumingia tellinoides)* are dirty white and have a distinct point at the rear shell. At their hinge, a spoonlike depression beneath the umbo protrudes into the inner shell.

HABITAT: Purplish and cancellate semeles live in sand banks off beaches out to moderate depths. Tellin and white Atlantic semeles prefer inlet areas and shallow bays open to the sea.

DID YOU KNOW? A semele's slash- or spoonlike hinge depression (chondrophore) bears a cushiony pad that springs the valves open when the animal's adductor muscles relax.

Tagelus (Short Razor) Clams, and Gray Pygmy-venus Clam

RELATIVES: Tagelus clams (family Solecurtidae) are related to semeles and tellins and are distant kin to pygmy-venus clams (family Veneridae).

IDENTIFYING FEATURES: These clams have central umbones and elongate shells that gape at each end.

Stout tagelus clams *(Tagelus plebeius)* have thick, lumpy shells with smooth growth lines. They are white, ivory, or light gray with a greenish periostracum on margins of freshly beached shells.

Purplish tagelus clams *(Tagelus divisus)* have smooth, thin shells that are tinted purple inside and out. A darker purple ray from the umbo marks a slightly raised internal rib. Small shells may have a covering of brown periostracum.

Gray pygmy-venus clams *(Timoclea grus)* have ribs crossed by growth lines and are cream or gray, often with a purple-brown streak covering the hind end.

HABITAT: These clams live in the sand or mud of shallow embayments. Stout tagelus clams prefer closed lagoons and purplish tagelus clams prefer bays open to the sea.

DID YOU KNOW? Tagelus clams live with only their siphons exposed and feed on suspended particles. Pygmy venus clams leave only their dark hind end exposed.

Stout tagelus, max 3.9 in (10 cm)

Purplish tagelus, max 1.6 in (4.0 cm)

Gray pygmy-venus clam, max 3/8 in (10 mm)

Calico clam, max 3.5 in (8.9 cm)

Sunray venus clam, max 6 in (15.2 cm)

Imperial venus clam, max 1.4 in (3.6 cm)

Venus Clams *(Calico, Sunray, and Imperial)*

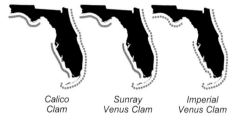

| Calico Clam | Sunray Venus Clam | Imperial Venus Clam |

RELATIVES: Venus clams are allied within the family Veneridae.

IDENTIFYING FEATURES: These venus clams have sturdy shells with forward-pointing umbones behind a conspicuous lunule (a rounded patch divided by the hinge line).

Calico clams *(Macrocallista maculata)* have smooth, creamy shells with blurry brown rectangles and smudges.

Sunray venus clams *(Macrocallista nimbosa)* have smooth, elongate shells that are purplish-brown with darker, narrow rays streaking from the umbo. Beach-worn shells may be bone-white.

Imperial venus clams *(Chione latilirata)* have their shells thickened by 5–9 concentric, chunky rolls. They are whitish, light gray, or mottled tan with a few blurry rays.

HABITAT: Calico clams and imperial venus clams live in sand off beaches out to moderate depths. Sunray venus clams live in the muddy sands of shallow bays.

DID YOU KNOW? The sunray venus is abundant in the Panhandle where it is popular with hungry gulls and other local chowder fans. The thick shell rolls of the imperial venus may help this shallow burrowing clam avoid predation by drilling gastropods.

Venus Clams *(Lady-in-waiting, Cross-barred, Pointed, and Princess)*

Lady-in-Waiting and Cross-barred Venus Clams | Pointed Venus Clam | Princess Venus Clam

Lady-in-waiting venus clam, max 1.6 in (4.1 cm)

IDENTIFYING FEATURES:

Lady-in-waiting venus clams (*Chione intapurpurea*) have strong concentric ridges that are serrated beneath on the hind end. Colors are cream, tan, or gray, often with brownish streaks.

Cross-barred venus clams (*Chione elevata*) have sharp, concentric ridges that cross radial riblets. Even beach-worn shells show a distinct cross-hatched look. Most are gray-white with a white or purple interior. South Florida shells have the most purple and may have colorful rays.

Cross-barred venus clam, max 1.3 in (3.3 cm)

Pointed venus clams (*Anomalocardia auberiana*) are rounded in front and tapered behind. They have strong concentric ridges and teeth at the inner margins. Colors are cream, tan, or gray with a white, brown, or purple interior. Some have faint blue-gray lines outside.

Princess venus clams (*Periglypta listeri*) have radial ribs and sharp ridges that form blades on the hind end.

Pointed venus clam, max 3/4 in (20 mm)

HABITAT: These clams live in shallow waters. Cross-barred and pointed venus clams are common in coastal lagoons.

DID YOU KNOW? Given the abundance of cross-barred venus shells, both fossil and recent, this species has been the most abundant clam in many Florida lagoons for about two million years.

Princess venus clam, max 3 in (7.5 cm)

Lightning venus clam, max 2 in (5.1 cm)

Disc dosinia, max 3 in (7.6 cm)

Elegant dosinia, max 3 in (7.6 cm)

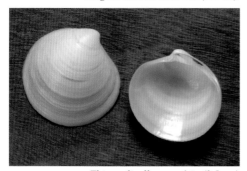

Thin cyclinella, max 1 in (2.5 cm)

Venus Clams *(Lightning, Disc Dosinia, Elegant Dosinia, and Thin Cyclinella)*

Lightning
Venus Clam

Disc and Elegant Dosinias
and Thin Cyclinella

IDENTIFYING FEATURES:

Lightning venus clams *(Pitar fulminatus)* have smooth concentric growth ridges and are thin-shelled for a venus. They are typically whitish with brown rays or dripping zigzags.

Disc dosinias *(Dosinia discus)* have ivory, circular shells with sharp forward-pointing umbones. They have fine concentric ridges too narrow for most folks to count without a hand lens.

Elegant dosinias *(Dosinia elegans)* are similar to disc dosinias, but their flattened concentric ridges are broad, easily seen, and readily felt.

Thin cyclinellas *(Cyclinella tenuis)* are flat white with fine but irregular growth lines. They are smaller and have thinner shells than the dosinias.

HABITAT: Each of these clams lives in sand or muddy sand from outside the surf zone out to moderate depths.

DID YOU KNOW? Dosinias have a hinge ligament strong enough to keep their valves attached long after their demise, surf tumble, and beaching. These events commonly follow encounters with predatory gastropods like moon snails. Many dosinia shells show the telltale holes from these meetings.

Quahogs *(Hard Clams)*

RELATIVES: Quahogs and other venus clams are in the family Veneridae.

IDENTIFYING FEATURES: Quahogs have thick shells with forward umbones, numerous concentric growth lines, and a heart-shaped lunule.

Southern quahog, max 5.9 in (15 cm)

Southern quahogs *(Mercenaria campechiensis)* have their mid-shell growth lines clearly visible, the largest of which are as wide as a pencil lead. They are gray outside with occasional purple zigzags and broad rays. Inside they are mostly white but may have hints of purple.

Northern quahogs *(Mercenaria mercenaria)* are similar to their southern cousins, but they differ in having finer growth lines that are smooth in the center of larger clams. Their inner margin tends to be deep purple.

HABITAT: These clams live in the muddy sands of shallow bays and lagoons.

Northern quahog, max 4.2 in (10.7 cm)

DID YOU KNOW? The southern quahog is our native Florida hard clam. Yankee quahogs are used by surf fishers for bait, which is why their broken bits are common on Florida beaches. Northern clams also have been commercially "seeded" in Florida lagoons, resulting in northern-southern hybrids. The genus *Mercenaria* translates to "payment," a reference to the wampum *(wampumpeg,* Algonquin for "valuable string of beads") made from the quahog's purple parts.

Northern quahog "bait clam" remnants

71

Angelwing, max 6.7 in (17 cm)

Campeche angelwing, max 5 in (12.7 cm)

Fallen-angelwing, max 2.8 in (7.1 cm)

False angelwing, max 2 in (5 cm)

Angelwings and False Angelwing

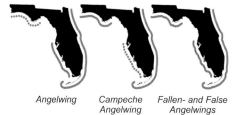

Angelwing   Campeche   Fallen- and False
         Angelwing   Angelwings

RELATIVES: False angelwings (family Petricolidae) are closer kin to venus clams than to angelwings (Pholadidae), which are allied with shipworms.

IDENTIFYING FEATURES: All have fragile, whitish, winglike shells with radial ribs.

Angelwings *(Cyrtopleura costata)* have a flared shell margin near the hinge that curves out at the umbo.

Campeche angelwings *(Pholas campechiensis)* have a flared margin in front of the hinge that curves out to cover the umbo. This membranous shell over the umbo is divided into several delicate compartments.

Fallen-angelwings (Atlantic mud-piddocks) *(Barnea truncata)* are stubby, with pronounced shell gapes both front and rear.

False angelwings *(Petricolaria pholadiformis)* have a simple hinge margin with 3 (left valve) or 2 (right valve) cardinal teeth.

HABITAT: All bore into muddy clay, peat, or rotten wood on the bottoms of open bays. Campeche angelwings also occur in shallow offshore clays.

DID YOU KNOW? True angelwings live with much of their soft parts outside their shells. The siphons of the fallen-angelwing extend 12 times its shell length.

Geoduck and Shipworm

Atlantic Geoduck   Shipworm

RELATIVES: Geoducks (family Hiatel-lidae) and shipworms (family Tere-dinidae) are bivalves related to angel-wings and piddocks.

IDENTIFYING FEATURES:

Atlantic geoducks *(Panopea bitrun-cata)* are beached as large, off-white, lumpy, oblong shells that clearly did not close without large gapes at either end.

Shipworms *(Bankia* spp. or *Teredo* spp.) are evident as snaking tunnels through beached driftwood. The tunnels of these bivalves are lined with white, fragile, shell material. These mollusks are worm-shaped, having small, wing-like, shell valves (generally deep in wood and not easily seen) in front and paddle-like "pallets" in the rear. *Teredo* has its pallets hollowed like a vase and *Bankia* has pallets composed of about 16 stacked funnels.

HABITAT: Geoducks live in burrows 4 ft (1.2 m) deep in muddy sand off beaches out to moderate depths. Shipworms live in submerged or drifting wood.

DID YOU KNOW? Geoducks are sel-dom beached until severe storms erode them from their deep burrows. Ship-worms tunnel for protection and feed by filtering outside water taken in by siphon tubes. Their shell valves close like jaws to grind wood and their rear pallets plug their tunnel to prevent dehydration.

Atlantic geoduck, max 9 in (23 cm)

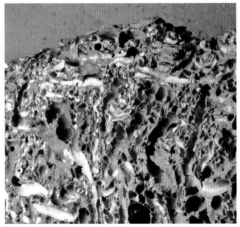

Shipworm tunnels, max 3/8 in (1 cm) diameter

Detail of shipworm tunnels

73

Circular Chinese-hat

Dwarf olive

Spotted pheasant

Snubnose corbula

Lunate crassinella

Atlantic nutclam

Atlantic abra

White strigilla

Miniature lucine

Many-line lucine

Itty-Bitty Shells

RELATIVES: Gastropods and bivalves

IDENTIFYING FEATURES: These shells are too small to be seen by folks on a casual stroll and are all less than about 1/4 inch (8 mm) as adults. The path into the amazing world of itty-bitty shells is traveled by those on their hands and knees. Peering into drift piles at the recent strand line will reveal many of this page's petite species in addition to miniature versions of the larger species shown on previous pages.

Dinky gastropods:

Circular Chinese-hat *(Calyptraea centralis),* all Florida beaches, family Calyptraeidae

Dwarf olive *(Olivella lactea),* all Florida beaches, family Olividae

Spotted pheasant *(Eulithidium affine),* SE Florida and Keys, family Trochidae

Wee bivalves:

Snubnose corbula *(Corbula caribaea),* all Florida beaches, family Corbulidae

Lunate crassinella *(Crassinella lunulata),* all Florida beaches, family Crassatellidae

Atlantic nutclam *(Nucula proxima),* all Florida beaches, family Nuculidae

Atlantic abra *(Abra aequalis)*, all Florida beaches, family Semelidae

White strigilla *(Strigilla mirabilis),* all Florida beaches, family Tellinidae

Miniature lucine *(Lucina amianta),* all Florida beaches, family Lucinidae

Many-line lucine *(Parvilucina multi-lineata),* all Florida beaches, family Lucinidae

Fossil Shells

Occasionally, beachcombers find fossil shells that were dumped along with inland fill material or that eroded from ancient deposits. These finds are most common where inlets cut through former beaches and where fill has been dumped from inland quarries.

A. *Mitra (Pleioptygma) heilprini,* family Mitridae

B. *Vasum horridum,* family Vasidae

C. *Cancellaria plagiostoma,* family Cancellariidae

D. *Fusinus caloosaensis,* family Fasciolariidae

E. *Conus adversarius,* family Conidae

F. *Fasciolaria apicina,* family Fasciolariidae

G. *Crucibulum multilineatum,* family Calyptraeidae

H. *Hanetia mengeana,* family Muricidae

I. *Eupleura intermedia,* family Muricidae

The seashells above have been extinct for roughly two million years.

J. *Nodipecten* sp., family Pectinidae, extinct 1–15 million years

K. *Nodipecten nodosus,* family Pectinidae, still with us (see page 52)

L. *Chione latilirata,* family Veneridae, still with us (see page 68).

M. *Eucrassatella speciosa,* family Crassatellidae, still with us.

N. *Chione cancellata,* family Veneridae, still with us (see page 69).

O. *Plicatula gibbosa,* family Pectinidae, still with us (see page 52).

Boreholes from bivalves (A) and gastropods (B)

Boring sponge perforations in a quahog

Polychaete worm grooves

Shell Wars (Shell Bioerosion)

Beached mollusk shells often bear clues to how they met their demise and who made use of them after their death. This evidence includes boreholes, perforations, and grooves.

Shells with single, circular **boreholes** were likely eaten by a predatory gastropod. Oyster drills *(Urosalpinx* spp.) leave a straight hole, whereas thick-lipped drills *(Eupleura caudata)* leave a slightly beveled hole. Shark's eye snails *(Neverita duplicata)* leave a countersunk, circular borehole that has an outer diameter about twice the inner diameter. Two tactics for hole-boring gastropods include edge drilling and umbo drilling. Drilling at the valve edge is fastest (because the shell is thinner) but is risky because closing valves could pinch the snail's proboscis. Umbo drilling is safer, but in the time it takes to bore through the thick umbo, a snail may have its prey stolen by a larger gastropod or become a meal itself (note the bored shark's eye in the top image).

Scattered **perforations** in a shell were likely made by boring sponges *(Cliona* spp.). These sponges partially acid-digest living and dead shells and invade them as living space.

Other animals that use shells as living space include polychaete worms like polydorids, which leave snaking **groove** marks. The router-tool indentations are made as the worm rasps with its bristled body aided by acids it secretes. Bivalve shells may also be penetrated by other bivalves like *Gastrochaena*, which leave oblong boreholes (A) in either shell or rock. This bean-shaped clam lives out its life within the pit it forms.

Shell Color Variation and Mollusk Bits and Pieces

Although some shells are most colorful in life, other shells turn a variety of colors after they die. These colors depend on the shell's afterlife experiences. Black shells were likely darkened by iron sulfide after burial in sulfurous muck. A beach speckled with numerous black shells indicates that the surf zone was once the lagoon behind the barrier island. Pink, rust, or brown are the colors most shells turn after decades of exposure to air. Although glossy white shells are probably recent, bone-white shells may be fossils. After millennia under ground and water, shells are slowly converted to the most stable form of calcium carbonate, calcite, the lime in limestone.

Colorful incongruous arks from the same beach

For every whole shell found on a beach there are thousands of bits and pieces. Some surf-worn shards have the clear distinguishing features of the original shell. Do you recognize these?

A. Operculum (trapdoor) from a **horse conch** *(Triplofusus giganteus)*

B. Giant tun shell *(Tonna galea)*

C. Queen helmet *(Cassis madagascariensis)*

D. Hawk-wing conch *(Strombus rainus)*

E. Lettered olive *(Oliva sayana)*

F. Shark's eye *(Neverita duplicata)*

G. Angelwing *(Cyrtopleura costata)*

H. Junonia *(Scaphella junonia)*

I. Nacre from a **penshell** *(Atrina spp.)*

J. Atlantic giant cockle *(Dinocardium robustum)*

K. Crown conch *(Melongena corona)*

Some readily recognizable shell fragments

Resources and Suggested Reading

Andrews, Jean. *A Field Guide to Shells of the Florida Coast.* Houston, TX: Gulf Publishing Company, 1994.

Morris, Percy A. *A Field Guide to Shells of the Atlantic and Gulf Coasts and the West Indies.* Boston, MA and New York, NY: Houghton Mifflin Company, 1973.

Rehder, Harald A. *National Audubon Society® Field Guide to North American Seashells.* New York, NY: Alfred A. Knopf, Inc., 1981.

www.jaxshells.org
www.shellmuseum.org

Index

Entries in **bold** indicate photos and illustrations.

Here are some other books from Pineapple Press on related topics. For a complete catalog, write to Pineapple Press, P.O. Box 3889, Sarasota, Florida 34230-3889, or call (800) 746-3275. Or visit our website at www.pineapplepress.com.

Florida's Living Beaches: A Guide for the Curious Beachcomber by Blair and Dawn Witherington. Florida has 1200 miles of coastline, almost 700 miles of which are sandy beaches. Exploring along those beaches offers encounters with myriads of plants, animals, minerals, and manmade objects—all are covered in this comprehensive guide with descriptive accounts of 822 items, 983 color images, and 431 maps. In addition to being an identification guide, the book reveals much of the wonder and mystery between dune and sea along Florida's coastline. (pb)

Dangerous Sea Life of the West Atlantic, Caribbean, and the Gulf of Mexico by Edwin S. Iversen and Renate H. Skinner. Learn how to avoid dangerous sea creatures—and how to administer first aid just in case you are unable to avoid them. Includes sections on species that bite, species that sting, species dangerous to eat, pests that harm swimmers, toxic mucus-secreting species, fish beak and processing injuries, and human/animal interactions at modern tourist attractions. (pb)

Common Coastal Birds of Florida and the Caribbean by David W. Nellis. 72 birds inhabit the zone where the sea meets the land in Florida and the islands south; this book shows the great variety of specialized behaviors developed by these birds to survive in this special habitat. Over 200 color photos show many features of these birds and their habits—never before have they been so fully illustrated. (hb, pb)

Seashore Plants of South Florida and the Caribbean by David W. Nellis. For backyard gardeners and serious naturalists alike, this book is a complete source for information about which plants grow best in nearshore environments. Describes characteristics of each plant, including form, flower and fruit date, habitat, and more. Discusses ornamental, medicinal, toxic, physical, edible, and ecological aspects of each plant. (pb)

Seasons of the Sea by Jay Humphreys. Illustrated by Jim Wilson. A fascinating but largely ignored world thrives within the coastal waters not more than a few yards or, at most, a few miles offshore in the vast Atlantic Ocean and the Gulf of Mexico. As the seasons change on land, so do they in the waters that surround Florida. In the summer in northeast Florida, for example, sea turtles come ashore to lay their eggs, and human activity is curtailed to protect the hatchlings when they make their run to the sea. (hb)

Florida Water Story by Peggy Sias Lantz and Wendy A. Hale. Introduces young readers to Florida's water systems and describes and illustrates many of the plants and animals that depend on these watery habitats. (hb)